ENERGY

DATE			

ENERGY:
A Conceptual Approach

Thomas E. Van Koevering
Nancy J. Sell

University of Wisconsin – Green Bay

Prentice-Hall Inc., Englewood Cliffs, New Jersey 07632

Library of Congress Cataloging in Publication Data

Van Koevering, Thomas E., (date)
    Energy: a conceptual approach.

    Includes bibliographies and index.
    1. Power resources. 2. Power (Mechanics)
I. Sell, Nancy Jean, (date). II. Title.
TJ163.2.V35 1986   333.79   85-3500
ISBN 0-13-277757-6

Editorial/production supervision
  and interior design: *Kathleen M. Lafferty*
Cover design: *Whitman Studio, Inc.*
Manufacturing buyer: *John B. Hall*
Cover illustration: *Pictures such as this one, a thermal infrared photograph, illustrate the relative heat loss that occurs from various portions
of a building. The film that was used to take this picture is similar to
ordinary film in that areas which receive the most radiation are lighter,
and the less-exposed areas are darker. In this case, the light areas indicate regions where there is greatest heat loss. It is necessary that the
specific colors observed be "assigned" by the special camera taking
the picture because the radiation that is actually used is not in the
range of visible light. (Courtesy of the U.S. Department of Energy's
American Museum of Science and Energy, operated by Science
Applications, Inc.)*

Printed in the United States of America

10 9 8 7 6 5 4 3 2 1

ISBN   0-13-277757-6   01

Prentice-Hall International (UK) Limited, *London*
Prentice-Hall of Australia Pty. Limited, *Sydney*
Prentice-Hall Canada Inc., *Toronto*
Prentice-Hall Hispanoamericana, S.A., *Mexico*
Prentice-Hall of India Private Limited, *New Delhi*
Prentice-Hall of Japan, Inc., *Tokyo*
Prentice-Hall of Southeast Asia Pte. Ltd., *Singapore*
Editora Prentice-Hall do Brasil, Ltda., *Rio de Janeiro*
Whitehall Books Limited, *Wellington, New Zealand*

To our families,
for their help, patience, and support.

Contents

Preface

Why another book on energy? Certainly data about the world energy dilemma are available in numerous publications. We do not claim to have uncovered any new or startling facts. In *Energy: A Conceptual Approach* we have attempted to relate the data about energy and energy use to a number of general concepts that can be understood by the general public. Data can change: There may be discoveries of new fuel sources or new technologies, government policies can shift, the economics of nations can vacillate, and sometimes we find that the original data were not only incomplete, they were wrong. Data alone seldom lead to new understandings because they are often selectively collected as evidence to support existing points of view. If people are going to understand the significance of new data as they become available, descriptions of the interrelatedness of energy and energy use to nearly every aspect of our lives are needed. In this book the descriptions of these relationships are sprinkled with enough data to provide the reader with a grasp of the importance of the concepts.

Energy often is thought to be only one of the many topics encountered in the process of learning about science. More specifically, energy is generally considered to be one of several components of a physics curriculum, along with mechanics, atomic structure, and so on. Restricting the study of energy to the application of basic physics principles does not do justice to the topic. Biology is more than learning names of organisms and their structures; learning chemistry involves much more than developing proficiency in balancing equations. The significance of energy transcends physics and other areas of science to include economics, geography, political science, and business. Learning about energy also requires delving into a study of the nature of people: What conditions in society cause people to use large amounts of energy and how people can be motivated to take a look at what they consider to be their "needs" and which of their "needs" may have to be reclassified as "wants." Energy does touch every aspect of our lives.

ACKNOWLEDGMENTS

We would like to express our appreciation to all who have assisted us in acquiring information used in the preparation of this book. We would especially like to thank Paul Wozniak, Wisconsin Public Service Corporation, and The American Museum of Science and Energy.

We have received invaluable assistance from our reviewers: Robert Knapp, Energy Systems Program, Evergreen State College; Robert H. Poel, Western Michigan University; P. Aarne Vesiland, Duke University; and Marshall J. Walker, Professor of Physics, The University of Connecticut. Special thanks goes to our production editor, Kathleen Lafferty, for her excellent advice, diligence, and ability to help us put together all the details.

We are indebted to our typists Jeanette Sell, Joy Phillips, Sharon Dhuey, and Hilda Wise for their efforts and mostly for their patience.

Thomas E. Van Koevering
Nancy J. Sell

CHAPTER 1

The Energy Shortage: Is the Problem Real?

People who express concern about energy believe that an energy shortage does exist and will continue to exist well into the next century. How do we know this statement is true? Public opinion polls, statements by U.S. Representatives and Senators, and authors of numerous articles support the contention that energy shortages are all contrived by oil companies and others as a method to justify dramatic increases in fuel prices.[1] How can we see through the rhetoric and arrive at any reasonable conclusions on this matter?

SHORTAGES

The term "shortage" requires some discussion. Webster[2] defines *shortage* as a noun referring to a lack or deficiency. An energy shortage can occur for any one of the several reasons. Some of the reasons are:

1. An energy shortage can occur for everybody whenever the immediate fuel supply cannot keep pace with the current demand. This

---

[1] Morris Goran, *Ten Lessons of the Energy Crisis* (Newtonville, Mass.: Environmental Design and Research Center, 1980), pp. 10-14.

[2] *Webster's New Collegiate Dictionary* (Springfield, Mass.: G. & C. Merriam Co., 1976), p. 1073.

has happened several times during the past decade. There simply is not enough fuel available at the right place to meet the immediate demand.

2. An energy shortage can occur for some people if the price of energy becomes too high. Some people may be asked to make extensive personal sacrifices in regard to their energy use because the price is too high. They may not be able to afford the cost of their basic heating, lighting, and transportation demands. They are experiencing a personal energy shortage even if there are not lines of people waiting at gasoline stations.

3. An energy shortage can occur when the energy that is available is not in a usable form. The earth receives enough solar energy each day to meet all the energy needs of all the people presently living on the earth. Fortunately, this energy is not highly concentrated, because plant and animal life could not survive if the temperature of the atmosphere and amount of high-energy radiation reaching the earth's surface were substantially increased. Unfortunately, the energy from the sun is not in a form in which it can be conveniently utilized directly by people to do the things that people want to do. Hence we can experience an energy shortage when actually we are continuously surrounded by more energy than we can use.

For people experiencing the shortage, knowing the reason for the shortage does not reduce the impact. However, the mental attitudes that people have will differ considerably if the resource is available to everyone rather than only to the affluent.

NONRENEWABLE ENERGY RESOURCES

Coal, petroleum, and natural gas, which account for well over 90% of our energy sources, are nonrenewable resources. That is, they will not be replaced in the foreseeable future. The production of these fuels by natural processes requires millions of years and the availability of these fuels for human use is measured in decades. These time frames are difficult to visualize together because it is difficult for the mind to grasp millions of years on the one hand and a few decades on the other.

Table 1-1 attempts to provide a single perspective. Suppose that we take all time since significant life first appeared on this planet and compress it into a hypothetical calendar that consists of 365 days. Significant life begins at 12:01 A.M., January 1, and we are located at 12:00 midnight on December 31. From the perspective given in the table, fossil fuels are produced in "months" and consumed in "seconds."

TABLE 1-1   Fossil Fuel Production and Use Time Scale

Hypothetical Year	Real Time Period
1 day	= 1,370,000 years
1 second	= 16 years
0.063 second	= 1 year

Hypothetical Year		Real Time Period
Jan. 1	Significant life	500,000,000 years ago
May 4 to Nov. 24	Oil formations	330,000,000-50,000,000 years ago
May 27 to Nov. 14	Coal formations	300,000,000-65,000,000 years ago
Nov. 17	Last of dinosaurs	60,000,000 years ago
Noon, Dec. 30	Last glacial age begins	2,000,000 years ago
Midnight, Dec. 30		1,370,000 years ago
Dec. 31		
11:47 P.M. (3 minutes ago)	End of Ice Age	12,000 years ago
11:57 and 54 seconds (126 seconds ago)	Time of Jesus of Nazareth	2000 years ago
11:59 and 29 seconds (31 seconds ago)	America discovered	500 years ago
11:59 and 52.7 seconds (13 seconds ago)	Industrial Age begins	200 years ago
11:59 and 52.7 seconds (7.7 seconds ago)	Oil discovered in Pennsylvania	1859
12:00 P.M.	Now	Present
2-6 seconds from now	Natural gas gone	32 years from now
4-6 seconds from now	Oil gone	64 years from now
19 seconds from now	Coal gone	300 years from now

If substantial new discoveries of oil, natural gas, and coal are discovered so that the fossil fuel supplies can actually last five times longer than the current projections, this would mean that we would be using fossil fuels at least 125 times faster than they are produced by natural processes. From the time perspectives just presented it is necessary to classify fossil fuels as nonrenewable resources.

ENERGY ECONOMICS

To obtain a more complete understanding of the problems associated with using a nonrenewable resource on a day-to-day basis, it is also necessary to consider some basic concepts of economics. The first concept to be considered is how the quantity of a nonrenewable resource varies over time. This is illustrated in Figure 1-1. Initially, very little use is

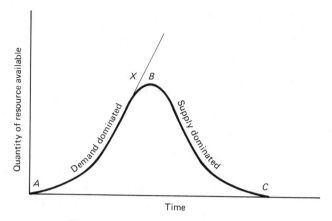

Figure 1-1   Resource use over time.

made of the resource because there is no need for it, and ultimately the resource is little used because it is nearly gone.

Why does the curve have this bell-shaped appearance, and what are the factors operating on the system? There is an increase in use of the resource on the left side of the curve (points A to B) because there is an increase in demand. People want it and supply is not a limiting factor. Among a number of reasons for the increase in demand may be the following: (1) a larger population, (2) the unavailability of a similar or competing resource, (3) the advancement of a technological system that requires the resource, (4) economic advantages due to a cheaper price, and (5) ease of handling and transport. In any case, the quantity of a resource used on a daily basis increases because people demand its use more and more. On the right side of the curve (points B to C), supply dictates how much will be available. As supplies decrease and we must look harder and longer for what is available, the amount of the resource that can be made available will diminish. Eventually, there is going to be a problem: The resource will be gone. Obviously, we are going to have to face up to reality and, we hope, do so long before we reach point C. We certainly cannot wait until the resource is exhausted before we start to look to alternatives, but when do we first begin to experience shortages?

Unfortunately, the shortage comes earlier than we might expect. In Figure 1-1 the demand-dominated portion of the curve has been extended as a straight line. This means that the rate of increase in demand, or percent growth each year, remains constant. Notice what happens at point X. The demand-dominated line separates from the curve. This indicates that demand has begun to exceed the supply available at that point in time. This then may be the simplest definition of a shortage: The demand exceeds the available supply. Also note that this problem occurs before even one-half of the resource has been used. This pattern correlates very well with what is actually happening with respect to our oil and natural gas reserves.[3]

A second economic concept that must be considered to have an appreciation for the worldwide implications of the energy dilemma is the diverse nature of the world economy. In developed countries such as the United States and Western Europe, the average annual income per person ranges from $3625 to $12,400, respectively, whereas in the developing countries (or as some refer to them, "the never to be developed countries") the average annual income per person is only about $610. These data clearly illustrate that there is a very significant difference in consumer buying power on the world market in these contrasted regions. A projection of these data using trends for the past 25 years illustrates that the future prospects for most of the world's poor people are very dim indeed. Consider the data given in Table 1-2. Not only is there presently an enormous difference in economic levels, but the outlook for the next several decades looks even more depressing. In the light of these data it is also easier to understand, from the reference frame of an educated person in a developing nation, why the efforts of the United States are not always appreciated in the way we might expect, especially when it comes to giving others advice about how they should use or prize their resources. As Kenneth Boulding, a noted economist, has so aptly stated: "One is particularly suspicious of good advice from the rich to the poor on how they can make the best use of their poverty."[4]

Let us return to the question posed by the chapter title, "The Energy Shortage: Is the Problem Real?" Note that if we take another look at the three definitions of shortages, we are left with three different answers. The first definition is that a shortage occurs when the immediate supply cannot keep pace with the current demand. There obviously have been times when energy supply shortages have existed. This could happen again because the largest energy producers are not all the largest energy consumers. Changes in political or economic conditions may again cause the suppliers to refuse to sell large amounts of energy to their major customers. For this reason, looking at our past and knowing that control of our energy supply does not rest entirely in our hands, there may well be an energy shortage "sometimes."

If we use the second definition—that shortages occur when the price is too high—there definitely is an energy shortage in many under-

[3] M. King Hubbert, "Energy Resources," *Resources and Man*, National Academy of Sciences-National Research Council (San Francisco, Calif.: W. H. Freeman and Company, Publishers, 1969), pp. 157-242.

[4] K. E. Boulding, "The Social System and the Energy Crisis," *Science*, Vol. 184, April 19, 1974, pp. 255-257.

TABLE 1-2   Estimates of Per Capita National Income
(expressed in U.S. dollars)

Region	1960	1978	1978 Constant Dollars*
World	520	2,310	1,049
Developed†	1,340	7,000	3,178
Underdeveloped	130	610	277
Africa	120	560	254
Ethiopia	47	91 (1975)	52
Asia (excluding Japan)	80	280	127
India	69	173	79
Pakistan	75	261	118
Japan	417	7,153	3,247
Latin America	76	269	122
Haiti	76	269	122
Europe	960	6,400	2,905
Netherlands	890	8,509	3,862
Spain	317	3,625	1,646
Switzerland	1,463	12,408	5,632
United Kingdom	1,261	4,955	2,249
United States	2,502	8,612	3,909
Middle East oil producers			
Kuwait	2,814 (1970)	11,431 (1975)	8,247 (1975)
Qatar	1,018 (1970)	9,929 (1975)	7,164 (1975)
United Arab Emirates	2,357 (1970)	16,665 (1978)	12,024 (1978)
Saudi Arabia	237 (1960)	6,089 (1975)	3,350 (1975)

*Represents real change using constant 1960 U.S. dollars.

†Developed = 33% of world population in 1960 and 27% of world population in 1978.

Source: Yearbook of National Accounts Statistics, Vol. II: International Tables (New York: United Nations, 1980), pp. 10-16.

developed countries. For people in countries where there has never been very much money available to buy energy, the price of energy has increased much faster than their incomes. In many parts of the world, "yes" is an appropriate response to the question, "Is there a shortage of energy?"

The abundance of solar energy available in most parts of the world means that there really is enough energy to do what people want done but that the energy is not in an appropriate form or is not sufficiently concentrated. At the present time, cost-effective technology is not being produced in sufficient quantity to utilize enough of the available solar energy to meet these needs. Yes, there can be a shortage in the midst of plenty.

QUESTIONS

1. How have people responded when a shortage has occurred because supply cannot keep up with demand? Does this behavior tend to relieve the shortage or to make it worse? What has the federal government threatened to do to restrict this behavior? Do you think this will work?

2. How have people often responded when they cannot meet their basic energy needs because the price is too high? What has the federal government done in some instances to assist people in these circumstances? Can you see any problems with this approach? Can you suggest any alternatives?

3. If continued use of the nonrenewable fossil fuels requires too stringent energy conservation measures to be enforced, consider the following:
   (a) What are your major energy uses at the present time?
   (b) How could you alter your life-style to reduce energy use and still continue to do many of the things you are doing now? (Example: By carpooling you could keep both your present home and your present job.)
   (c) What are some drastic steps you might consider if energy were in very short supply? (Example: Live in a smaller home, move closer to your work, etc.)

4. Cite examples of commodities that have progressed through the resource curve described in Figure 1-1. When were these commodities first used, when did they peak in use, when were they almost entirely replaced by something else, and why were they replaced? (Example: The use of horses on the North American continent.)

5. In Table 1-2, the percentage of people living in the developed world is shown to have decreased from 1960 to 1978. Give as many possible reasons for this as you can. Which reason do you believe provides the best explanation?

6. By what percentage has the U.S. dollar inflated in value from 1960 to 1978?

7. From Table 1-2, which country has had the greatest percentage increase in real buying power from 1960 to 1978? Which country has had the smallest percentage increase in real buying power from 1960 to 1978?

CHAPTER 2

The Nature of Energy
and How Nature Uses It

The universe is composed of matter and energy: Matter is the substance of the universe and energy is the mover of that substance. Every activity requires energy. This concept is a reality for every organism on earth. Plant life in any region is successful only if it is adapted to acquiring the energy it needs for survival. Near the north or south pole, this means acquiring the limited amounts of solar energy available over brief periods of time in the summer. Plants in tropical forests, which are provided with an abundance of sunlight, must compete with a wide variety of other plants for their "place in the sun." Animals, with some people as notable exceptions, spend most of their waking hours looking for food, and many are required to migrate long distances each year in order to sustain themselves. Obviously, energy is of great importance to all life—but what is it?

THE CHARACTERISTICS OF ENERGY

Energy is defined most easily in terms of what happens when it is used. Although energy is available in several forms—mechanical, heat, light, sound, electrical, magnetic, and atomic energy—it is measured directly only as mechanical or heat energy.

Mechanical energy is the energy associated with moving things—

those that move in place, such as a rotating pulley, and those that move from place to place, such as atoms and astronauts. Mechanical energy is often defined as the ability to do work. Work, in turn, requires that a force be exerted and, as a consequence of this force, an object is moved a certain distance. The amount of force that is required to move the object will depend on the mass of the object and the magnitude of the forces of friction and/or gravity that must be overcome. These relationships are given in the equations in Appendix A.

Heat energy was originally given quantitative values by defining the basic units of calories and British thermal units (Btu) in terms of a given quantity of water and the amount of thermal energy required to raise or lower its temperature a specific amount. However, the amount of heat needed to change the temperature of water 1.0 degree Celsius (°C) is not the same at all temperature ranges, and even varies over the 1.0°C temperature span, so it is difficult to obtain a precise measurement. This problem was solved by defining the calorie as being equivalent to 4.18 joules (J) of mechanical energy.

Another way of classifying energy is to consider it as either potential energy or kinetic energy. *Potential energy* is stored energy. There is a "potential" for doing work. Food, batteries, and objects located on high places have the potential for doing work. *Kinetic energy* is a measure of the work that is actually being done when something is moving. When a certain quantity of kinetic energy is obtained, an equal quantity of potential energy has been lost. The potential energy stored in gasoline is lost when this fuel is burned and a car travels a certain distance.

Energy can perhaps best be understood by comparing the amounts of kinetic and potential energy associated with familiar places, objects, and activities. Such a list is presented in Table 2-1.

ENERGY USE IN THE NATURAL SYSTEM

When we speak of an energy dilemma or energy crisis, it should be noted that this problem exists only for human beings and those domestic animals that are highly dependent upon people for survival. Although fossil fuels originated as a result of activities in the natural system, today human beings are the only creatures that make use of fossil fuels. It is interesting and instructive to compare how a natural system (one not dominated by human activity) and a system dominated by people relate to energy.

From the example in Chapter 1 of the "mythical year," which illustrates the relative time periods involved in the production and use of fossil fuel reserves, another significant observation can be made. The

TABLE 2-1   Kinetic and Potential Energies of Selected Objects

Object	Potential Energy (kJ)*
A fly (1 g) sitting on a table 3 ft above the floor	9×10^{-6}
An 8-lb (3.63-kg) bowling ball resting on a table 3 feet above the floor	0.0325
A 16-lb (7.26-kg) bowling ball resting on a table 3 ft above the floor	0.0650
A 3000-lb (1364-kg) automobile setting on a hoist in a garage 6 ft above the floor	24.4

Activity	Kinetic Energy (kJ)
An oxygen molecule traveling 2 miles per second (7200 mph)	5.5×10^{-19}
A fly (1 g) traveling 25 miles per hour	1.25×10^{-4}
A baseball (140 g) traveling 90 miles per hour	0.116
A bowling ball (7.25 kg) traveling 20 miles per hour	0.576
An automobile (1364 kg) traveling 55 miles per hour	412
An automobile (1364 kg) traveling 90 miles per hour	1104
All the water falling from Niagara Falls in 1 second	7000×10^3
Energy necessary to sustain an adult person for 1 day	10.5×10^3

*1 kJ = 1000 J.

natural system has been functioning for more than 500 million years on every part of the earth under a variety of conditions. Obviously, this system has achieved a balance between the amount of energy available and the quantity of life that can be supported. With a record that spans countless millenia, the natural system has developed credibility in utilizing energy that merits our attention. A look at the differences between how a natural system functions and how a system functions that is dominated by people is quite illuminating. These differences are summarized in the following four examples.

1. *In the natural system survival is fundamental.* This is true not only for each organism on an individual basis, but collectively for each species and ultimately the entire ecosystem upon which each of the individuals depends for survival. If the actions of any of the residents threatens their survival, ultimately that form of creature may disappear. No matter how attractive a new feature may be or how benevolent an

activity may be, if this places a species at a significant disadvantage in its ability to compete for food and other things vital to life, it will eventually become extinct. Human beings often consider themselves as distinct entities from the mass of living organisms that surround them. This is a perfectly acceptable philosophical position as long as human activity within the natural system does not compromise our chances of long-term survival. Using energy at a rate that exceeds the rate of availability can definitely put into jeopardy the long-term future of a large human population.

2. *Energy is the basis of life and wealth on our planet.* For every activity there is a limiting factor. A childish spree in a candy store may be limited by finances, the physiological constitution of the participant, or even the quantity of candy available. Ultimately, the activity will cease. Every piece of matter on this planet is recycled. The atoms that make up our bodies were, have always been, and will always continue to be part of this planet. This principle applies to everything that we encounter. The quantity, such as too much or too little water, or the quality, such as dirty air, may not be to our liking, but all we will ever have is already here. This is also true for our energy, which arrives as a gift each morning from outer space. Energy is available much as the manna was in biblical times. Like the manna, there is enough for the day and it is difficult to store.

Natural systems, as well as those dominated by people, must live on an energy budget. Natural systems, however, survive basically on the day-to-day supply, so they are not able to overspend as people often do.

We often base our economics on a standard established by tradition, such as gold, or products with immediate and often short-term usefulness, such as a source of cheap appliances. But times change. The value of gold is based on a perception of the relative value of world currencies. Gold, in itself, has very few specific uses. It does not directly sustain life. Energy makes the system go. A better measure of the wealth of a nation might be an evaluation of its long-term energy needs and energy resources. The world is just beginning to awaken to this reality.

3. *Progress in a natural system means better use—not more use—of energy.* Competition for energy is always significant in the natural system. In tropical areas sunlight may be abundant but plants are surrounded by competitive neighbors, trees with tall trunks, or other plants with large leaves. It is easy to be left in the shadows. A plant either adjusts so that it can live with the competition or it disappears. Some plants, particularly those living in frigid areas, have little competition from neighbors, but if they hope to survive their yearly energy crisis, the long cold winter, they must engage in survival tactics that are just as serious. Progress in nature means making better use of what is avail-

able. This progress is measured in small but deliberate steps. A few more grams of sustained plant or animal life (biomass) per square meter is progress indeed.

People's definitions of progress tend to lean toward quantity rather than quality. We measure progress in terms of growth. "How much" often outranks "how well." Time, however, favors quality, because quality survives. We will soon have to face this issue in terms of our recent record of energy consumption. Time is not on our side.

4. *Natural systems have carrying capacities.* Any component of a natural system adjusts to a shortage of one or more of life's essentials by migrating to a new territory, whether temporarily or permanently; reducing its birthrate; increasing the death rate; or by all these measures. The intentions of people, as revealed by their activities, illustrate that we either believe we can overcome our problems caused by an abundance of people by mechanisms not found in nature, or that the laws of nature somehow do not apply to us.

During the nineteenth century, migration was a significant factor in easing population pressures in Europe. That option has, for the most part, disappeared. The death rate in many parts of the world has been decreasing steadily due to progress in combating many diseases. This is particularly true in terms of infant mortality. There have been some small reductions in the worldwide birthrate, but given a little extra time, a yearly growth rate of 1.6% will accomplish the same ends as a growth rate of 1.9%. Each hour there are 20,000 births and 10,000 deaths. The consequences of this are a new population the size of Chicago every month and an additional U.S. population added to the world every 3 years. Perhaps the most important factor is that 70% of the world's adults and 80% of the world's children now make up that part of the world that we often refer to as the "have-nots."

ENERGY STORAGE ON THE EARTH

The amount of radiant solar energy the earth obtains is equal to the amount of radiant heat energy that the earth loses to the empty space that surrounds us. If there were no mechanisms available on earth for using the incoming solar energy, this planet could not support life. The two primary mechanisms are the hydrologic cycle and food chains.

One energy relationship in the *hydrologic cycle*—the latent heat process that accounts for part of the surface energy balance on the earth—is discussed in Chapter 3. There is also another aspect of the hydrologic cycle that involves energy, and it has been of particular interest to people. When water evaporates from the surface of the oceans and falls as precipitation on the land, it travels from a lower elevation to a higher elevation. This gives the water potential energy. When the water returns to the oceans, this energy is released.

The landscape near a mature river often reveals evidence of the amount of work that moving water can do. The Grand Canyon provides an excellent example. People have utilized this energy by building dams or using existing waterfalls to turn waterwheels or turbines; the products have ranged from cracked grain to electric power.

The Martian landscape in Figure 2-1 illustrates that water power is not a likely source of energy for people who plan to visit that planet because much of the landscape is nearly flat. It can also be noted from the photograph that Mars does not have a most important solar energy collector: plants.

Let's assume that we have a solar energy budget of 1,000,000 units of energy. On a worldwide average about 700 of these units will be captured by plants. The efficiency of the system can be increased tenfold by restricting our 1,000,000 solar energy units to those regions of the earth that are covered by plants. Now 7000 units can be captured by plants in a natural system and up to 10,000 units captured by plants if the region is being farmed. Plants on farmland do a better job of capturing sunlight because (1) the plants are all the same size; (2) the land is generally better than average, so the plants will grow better; and (3) people often add fertilizer to the soil.

Plants, by using up to 10,000 units of solar energy from the original 1,000,000 units, are capable of supporting all life on earth. The chemi-

Figure 2-1   Landscape of Mars.

cal reaction whereby plants utilize solar energy is called *photosynthesis*
and is described by the general equation

$$\text{carbon dioxide} + \text{water} + \text{sunlight} \longrightarrow \text{sugar} + \text{oxygen}$$

There are many other products in addition to simple sugars ($C_6H_{12}O_6$)
produced by plants because plants can directly supply all the nutrients
necessary for human life.

Photosynthesis is essential for plants because plants need energy in
order to live. Plants use their energy in the following ways:

25% for respiration
35% for growth and maintenance
35% for reproduction
5% removed by plant eaters

Respiration is the process used by all plants and animals to obtain
the energy they need to carry on the functions of their life. Respiration
is the opposite of photosynthesis, as the following equation indicates:

$$\text{sugar} + \text{oxygen} \longrightarrow \text{carbon dioxide} + \text{water} + \text{energy}$$

Many of the plants or parts of plants that are consumed by animals
generally are about 10% efficient in transporting energy to that animal.
Referring back to our energy budget of 1,000,000 solar energy units, a
maximum of 10,000 units are captured by plants and used to produce
food, while 990,000 units are radiated back into space as low-grade heat.
The plant can now pass about 1000 energy units along to a plant eater,
which leaves 9000 units of additional energy to be radiated back into
space as low-grade heat.

When the plant (or an animal) dies, all the solar energy stored in
the plant will be utilized by organisms called *decomposers*. Decomposers
may be insects of various forms or they may be single-celled bacteria
and fungi. These creatures, in turn, return the energy they have obtained
back to space as radiated low-grade heat.

What has been described so far is often called a *food chain*: plants
→ plant eaters → decomposers. The energy efficiency of the system can
be increased by adding more links to the chain. The plant eater can
pass along about 10% of its energy when it is consumed by an animal
eater (a carnivore) rather than by decomposers. Animal eaters, in turn,
can be consumed by other animal eaters, but in each step about 90%
of the remaining energy is lost.

Energy utilization in land-based systems is based primarily on the
numbers of plants and plant eaters. Energy utilization in water-based
systems is based primarily on the number of animal eaters: little fish

being eaten by bigger fish, and so on. An example of a land-based food chain follows:

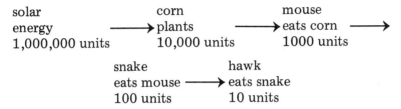

It is not difficult to see why food chains do not have 10 or 20 links: There is not enough energy available to sustain the members that would be included near the end of the chain. One can also speculate as to how many mice must be consumed to support one hawk. The natural system becomes more stable when food chains are expanded to become *food webs*. This means that each component depends on more than one source of food and in turn serves as food for more than one kind of creature. Mice can eat oats and wheat in addition to corn, and hawks can eat mice directly. The way that people use and manipulate food chains is explored in greater detail in Chapter 5.

BASIC ENERGY LAWS

The first basic *law of energy*, or *law of thermodynamics*, states that energy cannot be created or destroyed; it may be changed from one form to another, but the total amount of energy in the universe never changes. This is often referred to as the *law of conservation of energy*. To the energy consumer this means that there is no way to multiply energy. You can never receive more energy from any device than that which entered it originally.

The sun produces energy by changing mass into energy through nuclear fusion reactions that occur at very high temperatures and pressures. This does not violate the law of conservation of energy because matter and energy are interchangeable. This equivalence was first described by Albert Einstein in his famous equation $E = mc^2$, where E is energy, m is mass, and c is the speed of light. The speed of light is a large number, 186,000 miles per second (3×10^8 meters per second), so c^2 will be a very large number. This means that even a small amount of mass, m, results in the production of very large amounts of energy equivalent to that mass.

When you are gambling with energy there is no way you can win, that is, obtain a net gain in energy from the system. Actually, the second basic law of energy demonstrates that not only can you not win, but you will never break even. The *second law*, simply stated, tells us that although energy is radiated in all directions from any object not at absolute zero, the net result is that energy always flows spontaneously from a region of high concentration (high temperature), for example, to a region of low energy concentration (low temperature). Therefore, energy can never be transferred from a cold object to a warm object unless energy is added from outside the system, such as the electrical energy used to operate a refrigerator. Another implication of the second law is that things naturally move from an orderly state to a disorderly state. (Anyone who has spent any time around small children will never argue with that statement.) Order is restored only by using outside energy. When energy transformations take place, such as chemical energy changed to electrical energy in a battery, some energy is always lost as heat. Our earlier discussion of the food chain illustrates how large amounts of energy are lost in the system.

Energy can make only a one-way trip through any system, whether it is a food chain or a steam engine. Conserving energy thus means making the best use of the energy that is available and prolonging the use of that energy as long as possible before it is ultimately lost as radiated heat in space.

A term that is often associated with conservation is *efficiency*. Simply put, we usually think something is efficient when we get a lot back compared to what we put in. The first law of thermodynamics tells us that no energy system can ever be more than 100% efficient; that is, we can never receive more than we put in. The second law of thermodynamics tells us that we can never have a system that transforms energy from one form to another and have the process be 100% efficient. The culprit is usually waste heat.

The efficiencies of two different systems are described in the following example. In the first situation we will consider a heating system that heats a home using hot water. In this example, the hot water enters the room at a temperature of 180°F and leaves the room at 90°F. One measure of the efficiency of the system is to compare the total energy available from the heat source with the actual amount of heat lost. A heating system that is 100% efficient would lose all its heat to the surroundings to be heated.

In order to use the change in temperature to represent a change in energy, Fahrenheit degrees must be changed to an absolute temperature scale such as the equivalent reading in degrees Kelvin [whose unit is the kelvin (K): 180°F = 355 K and 90°F = 305 K]. The efficiency of the system can be calculated using the relationship

$$\text{efficiency} = \left(1 - \frac{\text{final temperature}}{\text{beginning temperature}}\right) \times 100\%$$

For the heating system the efficiency would be

$$\text{efficiency} = \left(1 - \frac{305}{355} \right) \times 100\% = 14\%$$

In the second example we will consider cooling a room using cold water. If the water entering is at $42°F$ and is at $60°F$ when it leaves, the equivalent absolute temperatures are 279 K and 289 K, respectively. Using the efficiency relationship that was just described, we have

$$\text{efficiency} = \left(1 - \frac{289}{279} \right) \times 100\% = -3.6\%$$

This represents a 3.6% change in the energy of the system. Under the conditions described, if the same quantity of water is involved, the system is nearly four times as efficient in heating the room as it is in cooling the room. From this example it is easy to see why air conditioners work harder than hot-water boilers to produce comfortable living conditions. It is also easy to see why the maximum amount of electrical energy demand for an electric power plant often occurs on a very hot day in summer.

It may appear that upon close inspection this example is in reality a comparison of "apples" and "oranges" because the energy "source" used to operate the air conditioner is invariably electricity, whereas it is not as likely that a household hot-water system would be run by electricity. However, electrical energy used in the home is not obtained directly from a naturally occurring electrical phenomenon; it is obtained by converting some other form of energy into electricity. If coal is used to heat the hot-water boiler and to generate electricity, the air conditioner loses for two reasons: (1) because the temperature change of the system is not very great, and (2) because less than 40% of the energy released by burning coal at the electrical generating plant is converted into electrical energy.

QUESTIONS

1. Why is the energy crisis a problem only for human beings?
2. Cite examples of how human beings use energy in ways that are similar to the ways in which other creatures use energy.
3. Cite examples of how human beings use energy in ways that are very different from the ways in which other creatures use energy.
4. How are human beings threatening their long-term future on the earth by the way they use energy?
5. Gold was cited as an example of a substance that has an artificial value. Give some other examples.

6. What are the two natural mechanisms that exist for storing solar energy on the earth? How do human beings use each of them?

7. Fossil fuels result from which process being interrupted—photosynthesis or respiration? How was it interrupted?

8. What is the difference between a food chain and a food web?

9. Why are food chains restricted to only a few steps?

10. Why is the amount of meat eaten per person in a particular country sometimes used as a crude measurement of the standard of living of that country?

11. If a person exercises rigorously while on a diet, does this not contradict either the first or second law of thermodynamics? Explain.

12. If, compared to a gas stove, an electric stove transfers more of the energy that enters the stove into heat energy that enters the food being cooked, why might a gas stove still be considered more energy efficient?

13. Why does cooling an automobile engine when it is running increase its operating efficiency?

14. Occasionally, an unenlightened person proposes that an electric generator be turned by an electric motor powered by some of the electricity from the generator. The electricity not used by the motor could be used for other purposes. What is wrong with this idea?

15. Why is cooling a room by $10°F$ using an air conditioner less energy efficient than heating a room by $10°F$ using a furnace?

CHAPTER 3

The Sun:
The Earth's Energy Resource

Nearly everyone recognizes that the sun, our nearest star, is an important source of energy for the earth. As other sources of energy increase in cost due to the realization of their finite nature, we are just beginning to consider the sun as a direct source of energy for human activity. We need to learn more about what the sun can and cannot do for us.

THE SUN AS A SOURCE OF ENERGY

The sun is important to us. People have known this for several thousand years, as evidenced by the numerous groups all over the world who are or have been characterized as "sun worshipers." Until recently, however, the sun's energy has been largely ignored by much of the technologically developed world as a direct energy source for human activity. We find ourselves in a circumstance somewhat similar to that of a tribe of natives encountered by a missionary for the first time. When the missionary learned that they worshipped the moon as their god, he asked why, if they were going to worship an object in the sky, they didn't worship the sun. They told him that the moon shines at night when they need the light, but the sun shines during the day when they can already see. We, like them, cannot afford to continue ignoring the obvious.

The sun is our primary energy source. As an energy producer it radiates more energy in 1 second than all the energy ever used by human beings on the earth. The sun consumes 564,000,000 tons of hydrogen each second in a nuclear fusion reaction which produces an interior temperature of 16 million degrees Celsius (25 million degrees Fahrenheit) and a pressure of 1 trillion pounds per square inch. These conditions allow four atoms of hydrogen to combine and form one atom of helium, with a loss of 0.7% of the mass of the hydrogen atoms. This mass is converted directly into energy. The sun's outer surface is at about 6000°C, but at this temperature 1 square meter radiates energy that is equivalent to the energy used by an 84,000-horsepower diesel train locomotive.

The sun appears to have been in its present form for more than 5 billion years and will probably continue for another 5 billion years before expanding to 60 times its present size. Then it will be classified as a red giant. All life on the earth at this time will be destroyed and turned to ashes. The sun will next contract to form a white-hot dwarf star no larger than the earth. Eventually, after many billion of years, it will cool off and "go out."

ELECTROMAGNETIC RADIATION

All the energetic activity of the sun would be meaningless to us—in fact, there would be no "us"—unless a mechanism existed for transporting this energy from the sun to Earth. It should not be surprising to learn that, indeed, such a mechanism does exist. It is called *electromagnetic radiation*.

The nature of electromagnetic waves was first recognized by James C. Maxwell, a British physicist, in the middle of the nineteenth century. The mathematical description of electromagnetic radiation is very complex, but the physical principles can be explained more easily. If an isolated electrically charged particle vibrates back and forth with a certain frequency, a magnetic field that surrounds the charged particle oscillates with the same frequency. A changing magnetic field produces a changing electric field, which in turn produces a changing magnetic field, and this process continues. These changing fields "radiate" from the source. If the speed of the radiation is too slow, the energy of the radiation will diminish. If the speed of the radiation is too fast, the energy of the radiation will attempt to build on itself. Neither of these conditions is consistent with the basic energy laws described in Chapter 2. The "just right" speed for radiation to continue indefinitely, millions of light-years in some cases, is about 186,000 miles per second (3×10^8 meters per second). That is equivalent to traveling approximately seven times around the earth in 1 second. The 93,000,000-mile trip from the sun to the earth requires about 8 minutes and 20 seconds.

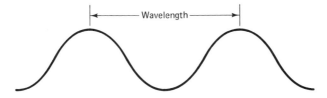

Frequency = number of peaks (waves) that pass a point each second
Velocity = frequency X wavelength = constant in a vacuum

Figure 3-1   Electromagnetic wavelengths.

Electromagnetic radiation has three important characteristics. All electromagnetic radiation travels in a vacuum at the speed of light, 186,000 miles per second. All radiation is not the same, however. There are obviously different colors of visible light, but visible light represents only a small fraction of the entire electromagnetic spectrum.

The distinguishing characteristics of electromagnetic radiations are their frequency and their wavelength. These characteristics are summarized in Figure 3-1 and Table 3-1.

The electromagnetic waves at different frequencies and wavelengths have different amounts of energy. As the frequency increases, the energy increases proportionately:

$$\text{radiant energy} = \text{a constant} \times \text{frequency}$$

$$\text{frequency} = \frac{\text{velocity}}{\text{wavelength}}$$

$$\text{radiant energy} = \frac{\text{a constant}}{\text{wavelength}}$$

TABLE 3-1   Frequencies and Wavelengths of Electromagnetic Radiation

Type of Wave	Frequency (per second)	Wavelength
Radio wave	1 million (10^6)	1000 feet
Microwaves	10 billion (10^{16})	1 inch
Heat (infrared)	10 trillion (10^{13})	$\frac{1}{1000}$ inch
Visible light	1000 trillion (10^{15})	$\frac{1}{100,000}$ inch
Ultraviolet	10,000 trillion (10^{16})	$\frac{1}{1\text{ million}}$ inch
X rays	100,000 trillion (10^{17})	$\frac{1}{10\text{ million}}$ inch
Gamma rays	1 million trillion (10^{18})	$\frac{1}{100\text{ million}}$ inch

We often encounter examples that illustrate that different forms of electromagnetic radiation have different properties. It is obvious that an X ray has greater pentrating power than any of the visible light waves because it has a higher frequency and a correspondingly shorter wavelength. Knowing that visible light can travel through glass and that infrared or heat waves cannot allows us to speculate correctly that infrared waves have a longer wave with lower frequencies than do visible light waves. Changes in the frequency of sound waves are detected by our ears as a change in pitch. Changes in the frequency of visible light waves are detected by our eyes as a change in color.

The fact that different forms of radiation have different impacts on matter should not be surprising. We feel heat and we see light. Our skin changes color in the presence of ultraviolet radiation. Radio waves and microwaves carry messages, and X rays and gamma rays have medical applications.

Electromagnetic radiation interacts physically with matter in several complex ways, but some simple examples can be cited. One of the most direct forms of interaction between an electromagnetic wave and a particle is to have the electromagnetic wave increase the velocity or speed of that particle. This is what happens to any substance that is exposed to direct sunlight. We can detect that the molecules of the substance are traveling faster if we are aware that the temperature of the object has increased.

All molecules rotate about one or more axes, and the range of frequencies of this rotation is in the same range as the frequency of microwave radiation. If the frequency of the microwave matches the rotation frequency of the rotating molecule, the radiation is absorbed and the molecule rotates at a higher frequency. This is particularly noticeable with water molecules in a microwave oven. Food with a high water content tends to heat up more rapidly than does food that contains only small amounts of water.

Atoms within molecules vibrate back and forth, sometimes acting as though the chemical bonds were springs. The frequency of these vibrations is in the same range as that of infrared radiation or heat waves. When the radiation frequency matches the frequency of two vibrating atoms, the radiant energy is absorbed. This principle can be used to identify substances utilizing an infrared light.

Visible light and ultraviolet light cause electrons to change energy levels within atoms. This can result in the chemical bonds between some atoms in a molecule being broken. The damaged molecules may be troublesome if they are part of a living organism. It is fortunate for life on the earth that much of the ultraviolet radiation radiated from the sun is absorbed in the upper atmosphere of the earth before it can be absorbed by plants and animals. Ultraviolet radiation can have dangerous effects, such as an increased incidence of cataracts and skin cancer in people.

The interaction between radiated energy and matter also occurs as a result of nuclear reactions; that is, changes within the nucleus of an atom. In human beings this radiation may be absorbed through the skin, but it can also be ingested directly into the body through the food we eat and the air we breathe. This closer contact greatly increases the number and severity of problems that can occur in the human body as a result of exposure to radiation that comes from nuclear reactions.

SOLAR ENERGY AND THE ATMOSPHERE

The earth is about 93,000,000 miles from the sun and the earth has a diameter of only 8000 miles. This represents a very small target in space for radiated solar energy. We receive about 1 part in 2 billion parts of all the energy radiated from the sun. The rate of delivery of even this much energy, however, amounts to nearly 180,000 trillion watts continuously striking the outer atmosphere. Fortunately, all this radiation does not strike the earth. If it did, the average temperature of the earth would be much higher than it is now. The interaction of solar energy with the atmosphere is shown in Figure 3-2.

About one-third of the energy from the sun is transmitted directly back into space from the earth's atmosphere. This part of energy is the light that is seen by earth satellites and a few human voyagers in space.

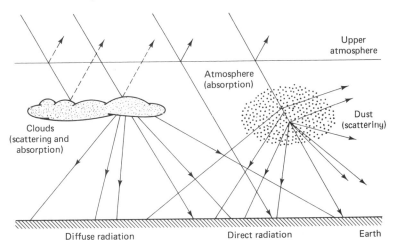

Figure 3-2 Solar radiation and the earth's atmosphere. (Courtesy of the U.S. Department of Energy's American Museum of Science and Energy, operated by Science Applications, Inc.)

The scattered energy consists primarily of radiation in the short-wave or blue end of the visible spectrum. This is why the sky is blue. At sunrise and sunset the sunlight has to travel a greater distance in the dense part of the atmosphere near the earth's surface; thus the scattering effect is much more extensive and only the long red waves are able to make the journey, which accounts for the red light we see at those times of the day.

THE EARTH AS AN ENERGY SOURCE

The earth is also a source of radiant energy. Any object that has a temperature above absolute zero, $-273°C$ or $-460°F$, radiates energy to its surroundings. The nature of the radiation, and its frequency and wavelength, depends on the temperature of the radiating surface. The sun, with a surface temperature of about $9000°F$, radiates energy that is primarily in the visible and ultraviolet range (short-wave radiation). The average temperature of the earth's surface is about $65°F$, and thus the radiant energy from this surface is much less energetic, with longer wavelengths. This radiation is in the infrared or heat range of the electromagnetic spectrum.

A principle that is paramount in understanding energy relationships on the earth is that the earth must radiate back into space as much energy as it receives from the sun. The average temperature of the earth has probably not changed by more than $10°F$ either way from $65°F$ in the last 1 billion years. Even changes this small resulted in expanded areas of either tropical or glacial conditions.

It should also be obvious, using our own familiarity with the various climatic regions of the earth, that solar radiation is not evenly distributed over the surface of the earth. The regions of the earth within $40°$ latitude of the equator receive more energy from the sun than is radiated back into space. The regions above $40°$ latitude radiate more energy than is received. If this condition continued in each locale, the temperature there would steadily increase or decrease. We know that this does not happen, so other processes must be involved.

An energy balance is maintained over the surface of the earth through three major mechanisms. The primary mechanism is that of air currents. Cold air moving away from the poles to below $40°$ latitude absorbs some of the "excess" radiation the earth has received in this warmer region. Warm winds from the tropics bring thermal energy to the more polar regions that are experiencing a net loss in energy. This process makes up for about 50% of the uneven distribution of solar energy on the earth's surface.

Ocean currents redistribute about 20% of the excess energy near

the equator and deliver it to the polar regions. It is interesting to con-
template how climates in many parts of the world would change if the
earth rotated in the opposite direction. The clockwise rotation of
major ocean currents in the northern hemisphere would rotate counter-
clockwise like those in the southern hemisphere, and vice versa. The
United Kingdom and Greenland would trade climates. The coastal
regions of Alaska would resemble present-day Siberia. Fortunately,
the direction of rotation of the earth will never be a debatable item at
the United Nations.

The remaining 30% of the energy that must be redistributed is
accounted for by the processes of evaporation and condensation of
water. These processes are a little more difficult to understand because
they do not involve warm air or warm water, where we can experience
their effect directly. When ice melts at $0°C$ and forms a liquid at $0°C$,
80 calories (cal) is required to produce this change for 1 gram (g) of
ice. This energy is needed to separate the molecules that are in the solid
crystal phase to those in the liquid phase, where they move somewhat
independently of each other. When liquid water evaporates to form
water vapor or boils to form steam, the molecules are much more widely
spaced and more independent of each other. The energy required to
bring about this change is at least 540 cal for each gram of water.

Most of the water vapor that forms precipitation originates within
$25°$ latitude of the equator. For every gram of water that evaporates
from this region, 540 cal of energy or more is removed from the surrounding water surface, which then becomes colder. If this water, which
is at the same temperature as the atmosphere, remains in the vapor
phase until it is transported to beyond $40°$ of latitude (about 2700
miles from the equator) and then condenses to form a water droplet in
a cloud, the 540 cal is added to the surrounding atmosphere at that
location. The net effect is the transport of 540 cal from near the
equator to a more polar region.

TAMPERING WITH THE ENERGY BALANCE

The natural system has been able to survive, and in some instances
thrive, by depending on a nearly constant average temperature for the
earth. With the advent of a large human population, and the food and
shelter demands they place on the natural system, the need for temperature stability has increased. It is interesting to note, however, that
the creatures that have the greatest dependence on appropriate amounts
of precipitation and appropriate temperature conditions for high productivity of food are the creatures that, by their own activities, are
threatening the system.

Threats to the energy balance of the earth by the activities of people can be placed in two general areas: the impacts of the greenhouse effect and of particulate matter in the atmosphere. The *greenhouse effect* relates to the phenomenon whereby short-wave or high-energy radiation such as sunlight travels through a medium such as the glass in a greenhouse and is absorbed by the contents of the greenhouse. The absorbed energy is reradiated as long-wave or low-energy radiation that cannot travel through the glass; hence the energy is trapped within the structure.

Carbon dioxide and water vapor found in the atmosphere surrounding the earth's surface are similar to the glass in the greenhouse. These molecules do not interfere with incoming short-wave radiation but they absorb the long-wave radition that is reradiated from the earth. This energy is, in turn, reradiated by the H_2O and CO_2, and some of it is radiated back to earth. The net effect is that the temperature of the atmosphere is about $18°F$ warmer than it would be without either of these gases in the atmosphere. As the amount of CO_2 or H_2O in the atmosphere changes, the temperature of the atmosphere also changes.

Carbon dioxide and water vapor molecules behave in a manner similar to that of a blanket on a bed. They are not a source of energy but they retard the radiation of energy. An extra blanket on the bed warms the occupant because the person's own body heat is not lost to the room as readily. The temperature of the air near the person increases. Eventually, however, a new balance point is reached and the rate of heat loss again must equal the rate at which it is produced. But the temperature near the person is now at the higher level.

The amount of water vapor in the atmosphere is highly variable depending on the local weather. Because of the large amounts of surface water on the earth, human activity has had very little impact on the total amount of water vapor that exists in the atmosphere. This is not true, however, for carbon dioxide.

Carbon dioxide is removed from the atmosphere by plants through the process of photosynthesis and replaced when the plant respires, is consumed, or decays (see Chapter 2). Fossil fuels were formed from plants that died but did not decay in the traditional sense. Some of the carbon dioxide removed from the air by photosynthesis was not replaced. Gradually, over millions of years, large amounts of carbon dioxide were removed from the atmosphere. The impact on a year-to-year basis would not be noticeable because fossil fuel formation is a slow and more or less continuous process.

In recent years, people have had an impact on the concentration of carbon dioxide in the atmosphere by burning large amounts of fossil fuels in a short period of time. When 1 ton of coal is burned, about

3 tons of carbon dioxide is produced. (Carbon dioxide consists of 27% carbon and 73% oxygen. Most of the oxygen comes from the atmosphere, which means that over two-thirds of the mass of the CO_2 produced came from the air.) The result has been an increase in CO_2 from about 280 parts per million by volume in 1880 to about 320 parts per million by volume in 1958. About one-third of the increase came as a result of people burning fossil fuels and about two-thirds by people clearing the forests and burning wood. The sharper rise in CO_2 levels since 1950 are attributed primarily to the burning of fossil fuels.[1]

The predictable change in atmospheric temperature due to the added CO_2 is probably about $+0.5°F$. Future possibilities with the use of even greater amounts of fossil fuels give rise to predictions of atmospheric temperature increases that range from a high of $10.5°F$ by the year 2000 to 8 or $9°F$ by the year 2150.[2]

Predictions are significant, however, only when the model being used also explains changes to date. The carbon dioxide model alone is inadequate to explain what has happened to the average atmospheric temperature for the past 100 years. From 1880 to 1940, the average global temperature rose about $0.5°F$ and then dropped slightly from 1940 to 1965 and has remained fairly constant since then. If CO_2 levels have been steadily increasing, why the drop in temperature between 1940 and 1965? When fossil fuels are burned (with the exception of natural gas), particulate matter, or *soot* as it is frequently called, is added to the atmosphere. The particles can cause water vapor to condense and form more clouds, which allows less solar energy to be absorbed by the atmosphere either directly or indirectly. Thus the atmosphere cools. We know that particulate matter can cool the atmosphere because large volcanic eruptions in the past have resulted in global cooling effects that have lasted for a year or more.

It is possible that large amounts of particulate material added to the atmosphere may be offsetting the impact of more CO_2 in the atmosphere. If this is indeed the case, an interesting situation presents itself. As long as the effects of the pollutants, CO_2 and particulates, compensate for each other, it is similar to a person attempting to balance on stilts that are each getting longer. If each stilt grows at the same rate as the other stilt, it is possible to maintain one's balance even if the consequences of a fall become more perilous. What happens if the concentration of one of the pollutants is suddenly decreased? Then we are left on top of one tall stilt.

[1] Minze Stuiver, "Atmospheric Carbon Dioxide and Carbon Reservoir Changes," *Science*, Vol. 199, 1978, pp. 253–258.
[2] S. Manabe and R. F. Wetherald, "The Effects of Doubling the CO_2 Concentration on Climate of a General Circulation Model," *Journal of Atmospheric Sciences*, Vol. 32, 1975, pp. 3–5.

The amount of particulate matter in the atmosphere will probably be significantly reduced in a very short period of time, perhaps just a few years, when we have no more coal to burn. This will leave us with a large amount of CO_2 in the atmosphere, which will increase the temperature of the atmosphere. The oceans will eventually consume the excess CO_2 in the atmosphere as it dissolves in the water and eventually forms solid calcium carbonate, but this process may take several hundred years.

The ultimate problem is one that people can create with respect to the temperature of the atmosphere. Energy expended from whatever source, for whatever reason, eventually is radiated as heat from the surface of the earth. This expended energy is added to the existing solar energy budget that must be reradiated back into space. This is similar to putting a hot water bottle under the blanket in our example; the bed heats up, and so will the atmosphere. Today the energy radiated from human activity is only about 0.01% of the energy we receive from the sun. If this value reaches 0.5 to 1.0%, it can begin to have a determining effect on the earth's climate.

POTENTIAL FOR CLIMATIC CHANGE

The magnitude of the temperature changes that have been described in the preceding section range from less than 0.05°F to a maximum of perhaps 5°F. The temperature at any given location may change by this amount in a matter of only an hour or two. Why, then, are some people so concerned about temperature changes that will neither freeze us to death nor produce conditions of unbearable heat? The answer is that an increase in the earth's average temperature of even 3°F can result in changes that affect people in two ways.

The popular notion is that if the earth's temperature increases, the glaciers will melt, which will increase the levels of oceans and cause flooding of coastal regions to elevations of 300 or more feet. (The highest point in Florida is only 180 ft above sea level.) This can happen and the results would be devastating because most of the world's population lives at elevations of less than 1000 ft above sea level. However, these events will require centuries to be completed.

The most notable short-range impact will be shifts in the major crop-growing regions of the world. If the major grain crops have to be grown in more northerly regions of the world, they will be grown on soil that is much less productive. Violent weather associated with rapid climatic changes could also sharply curtail crop production. The ever-expanding world population cannot tolerate large fluctuations in grain production. The 10 to 12 million people who now die directly from starvation each year could represent only a small percentage of yearly starvation deaths in the future if these significant climatic changes occur.

PREDICTING CLIMATIC CHANGE

The last 1,000,000 years has been characterized by at least eight major glacial periods, each lasting about 100,000 years, with intermittent periods of 10,000 years or more of warmer temperatures. In between the major ice ages have been times of mini ice periods, each lasting two or more centuries. The temperature variations associated with glacial periods are of the order of $10°$F and the mini glacial periods resulted in fluctuations of 1 or $2°$F. These events all transpired without the influence of human activity and are for the most part unexplained phenomena. We are part of a climatic roller coaster over which we have no control, and yet because of our ever-increasing needs for food, we have to focus on bumps in the track.

A striking anomaly thus appears. People characterize themselves as being the creatures most capable of adapting to wide-ranging climatic conditions. This is very true when the numbers of people are small. Because of the need for more and more food, however, large populations of human beings have become the creatures most dependent on the impossible: a world that does not change.

QUESTIONS

1. Why does 1 lb of fuel in a nuclear reaction produce thousands of times more energy than 1 lb of fuel in any chemical reaction?
2. When the *Mariner* spacecraft lands on Mars 35,000,000 miles from Earth, how long would it take a message from Earth to reach the spacecraft and a return response to be sent back to earth?
3. What color of light from the sun is most easily scattered by the atmosphere? What two observations can be used to answer this question?
4. What is one way in which people can easily determine that the atmosphere of the earth is heated primarily by radiation from the earth rather than being heated directly by solar radiation?
5. How do we know that the earth radiates into space as many kilojoules of radiation as it receives from the sun?
6. How can precipitation that falls in Canada transport energy to Canada from the Gulf of Mexico?
7. How have people increased the amount of carbon dioxide in the earth's atmosphere?
8. How does particulate matter affect the temperature of the atmosphere?
9. What would be some of the significant impacts on people if the average temperature of the atmosphere were to increase significantly?

CHAPTER 4

Energy Use:
How Did We Get Here
from There?

A convenient unit of energy for people is the amount of energy necessary for a person's body to function in a normal way. The number of kilocalories available to people from food depends on where people live. The hungry world, which comprises about 72% of the world's population, averages about 2100 kilocalories (kcal) of energy from food per person each day. The satisfied world averages about 3100 kcal of energy from food per person each day. Using the energy equivalences given in Appendix A, 3000 kcal per day per person is equal to 145 J per second or 145 watts (W). This means that the energy consumption of a person is about the same as the energy consumed by a 150-W light bulb that burns 24 hours a day. It is amazing what people have been able to accomplish at a rate of 150 W. If people were able to function using electrical energy directly (at a cost of 5 cents per kilowatt-hour), our weekly "grocery bill" would be about $1.25.

INCREASES IN ENERGY CONSUMPTION

When people began to utilize fire, wind, water, and domestic animals, the amount of energy available to each person increased to 5 or 10 times over the original base value of 3000 kcal. All these changes occurred more than 2000 years ago.

The first extensive use of fossil fuels occurred about A.D. 1200 in

England. This was the first time people began to use solar energy that had been stored for millions of years. The energy bank was now in business.

Coal was used initially when there was a shortage of firewood in England. In the fourteenth century the 40,000 inhabitants of London had used nearly all the firewood that was within easy reach. The high price of carrying firewood over the land gave way to cheap water transportation of coal from Newcastle. Then, during the 100 years following the first plague in 1348, the population of England was reduced to 60% of its original level. Wood was then plentiful until about 1600, when inflated prices again made coal a cheaper alternative. The coal market was here to stay.[1]

Water power was also used as a major source for energy during the eighteenth century. The steam engine, however, changed things by the early nineteenth century. Fuel for the steam engine could be transported to places where water power was not available. Industries could be much more widely dispersed. The next innovation in energy consumption came about 1870 with the development of the electric generator. Now energy, rather than fuel, could be transported. Electricity also allowed for extensive lighting, which meant that factories could now be productive 24 hours a day. Cities began to grow more rapidly than they had in the past.

The emergence of the internal combustion engine in the early twentieth century produced a market for petroleum. Although available since before the Civil War, the original use for petroleum was as a source of fuel for lamps, replacing more expensive whale oil. Automobiles created an entirely different demand for "black gold" (hardly gold, when it was selling for 10 to 15 cents a barrel at the turn of the century). Today, we have reached the point where the life-style of each person in the United States is capable of utilizing 75 to 100 times the original basic unit of 3000 kcal that is necessary for the body to function.

It is necessary to examine who is using 75 to 100 times as much energy as the caveman, because there are vast differences in energy use around the world. Figure 4-1 illustrates this difference by assigning a servant equivalent for each 3000 kcal per day used by people. The United States is given a value of 80, which means that the average person in the United States uses 80 times the 3000-kcal minimum each day. This is a phenomenal amount of energy. With less than 6% of the world population, the United States consumes 34% of the world's energy.

The world average is 12.9, but this is a very difficult number to conceptualize. If the United States were to reduce its energy consumption by 85%, from 80 to 13, to equal the world average, the results

[1] William TeBarke, "Air Pollution and Fuel Crisis in Pre-Industrial London, 1250–1650," *Technology and Culture* (University of Chicago Press), July 1975.

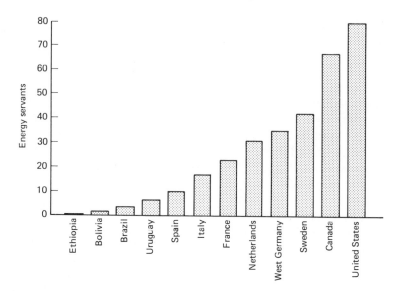

Figure 4-1 World energy use using servant equivalent units. (Courtesy of the U.S. Department of Energy.)

would be devastating. If this occurred, all our energy would have to be used to maintain our agricultural system. It is, however, possible to enjoy a life-style that is somewhat comparable to our own and use considerably less energy. The people in Western Europe do not have a "caves and candles" society, yet they use about one-half of the energy per person that we do.

For the person living in Ethiopia or Bolivia, 12.9 energy servant equivalents would be a dream come true. In Ethiopia, five people together have the equivalent of one additional person working for them. This energy poverty is especially harsh on the dependent population, the old and the young who cannot take care of themselves. There just is not enough energy to go around. A more comprehensive list of the relative amounts of energy used by several other countries is included in Appendix B.

EXPONENTIAL GROWTH

The important aspects of energy use that have been considered so far are (1) how we obtain our energy from the natural system, (2) how energy sources have changed over time, (3) the increased individual energy use, and (4) the discrepancies in energy use on a worldwide scale. One additional aspect of energy use requires further explanation at this point—the characteristics of growth patterns. Growth patterns can be divided into two general categories: arithmetic and geometric. An *arith-*

metic growth pattern exists when a constant amount is added at given intervals of time. As an example, if a farmer raises 10,000 bushels of corn one year, and then increases the amount of corn produced by 1000 bushels each year for 10 years, a constant value is being added to the initial value. Projections into the future represent the same rate of growth no matter what time frame is used. In 20 years the farmer will be producing 30,000 bushels per year and in 40 years, if the growth rate is constant, the yield will be 50,000 bushels per year.

A *geometric growth pattern* emerges when growth occurs by adding a constant percentage of what occurred the previous year. If this farmer were able to increase production by 10% each year, rather than the constant 1000 bushels each year, this production would be an example of a geometric growth pattern. The data compiled for crop production for the first 40 years are given in Table 4-1. The difference in the data generated occurs because in the second instance the base becomes larger each time, so 10% of each successive number results in a larger value again being added to the new base. In a 10-year span we can see that a 10% growth rate results in the production of substantially more corn than the amount that results from adding a constant 10% of the initial value. What about the future? The geometric growth pattern resulted in a doubling of production around the seventh year. After 14 years the 20,000 will have doubled again to become 40,000, and after 21 years the 40,000 will have increased to 80,000. The production doubles every 7 years. In 70 years, 10 doubling times, the production will be 2^7, or 128, times the original value or 1,280,000 bushels. This compares with 80,000 bushels if the growth pattern were arithmetic. A graphical representation of these data is given in Figure 4-2.

TABLE 4-1   Example of Arithmetic and Geometric Growth Patterns

Year	Constant Addition Each Year (bushels)	10% Increase Each Year (bushels)
0 (the reference year)	10,000	10,000
1	11,000	11,000
2	12,000	12,100
3	13,000	13,310
4	14,000	14,641
5	15,000	16,105
6	16,000	17,716
7	17,000	19,487
8	18,000	21,436
9	19,000	23,579
10	20,000	25,937
20	30,000	67,275
30	40,000	174,500
40	50,000	452,600

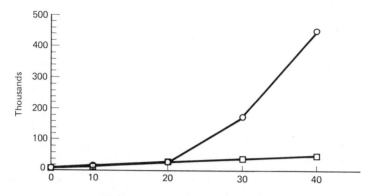

Figure 4-2 Exponential growth.

How does all of this relate to energy use? Figures 4-3 and 4-4 illustrate energy use patterns in the United States. The shape of the curves in these figures illustrates that population growth and energy use are both geometric growth patterns. The composite of both curves describes an interesting phenomenon, as shown in Figure 4-5. Energy growth in the United States from 1850 until 1900 increased at the same rate as the population. More people required more energy, but the amount of energy used per person during that time period remained relatively constant. Since 1900, however, energy consumption has increased at a faster rate than the population has increased. The amount of energy used per person was going up. The differences amounted to 1% per year growth for population and about 4% per year growth for energy use. It means that the population doubled every 70 years and energy use doubled

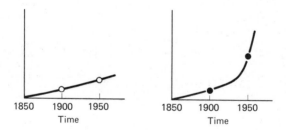

Figure 4-3 Population.   **Figure 4-4**   Energy use.

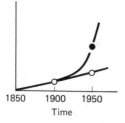

Figure 4-5 Energy use and population.

every 17 years. A comparison between 1900 and 1970 means that there were two Americans in 1970 for every one in 1900 and that each American in 1970 used eight times as much energy as his or her counterpart in 1900. The total energy consumption in 1970 was 16 times greater than the total in 1900.

The mathematical relationship for predicting doubling time is a simple one:

$$\text{doubling time (years)} = \frac{70}{\text{percent growth each year}}$$

An easy way to visualize this relationship is to picture having $100 in a savings account. If this account earns 1% interest each year and the interest is compounded (that is, added to the principal), in 70 years the original amount of $100 will double; if the account earns 2% a year, it will double in 35 years; and so on.

Using this relationship, the growth implications for energy use in the future can be frightening. During the 1960s the rate of increase in electric power consumption was about 7% a year. Thus electric power production was doubling every 10 years. For every electrical generating plant in operation in 1960, two were needed in 1970. If this figure is extrapolated, four plants were required in 1980, eight will be required in 1990, and 16 in the year 2000! Fortunately, things changed in the 1970s. Energy conservation made a notable impact on energy consumption at the time, but more important, the growth rates of energy consumptions was sharply curtailed.

Following is an example of an item that has been doubling in cost about every 7 years. In 1950 adult movie prices were 25 cents, by 1957 this cost had increased to 50 cents, to $1 by 1964, to $2 in 1971, and to about $4 in 1978. If this trend continues, movies should cost $32 in 2001.

The implications of geometric growth patterns on the future availability of fossil fuels are considered as each fuel is discussed in the following chapters. A few general examples will be cited here. The data in Table 4-2 illustrate the impact of increasing consumption of a nonre-

TABLE 4-2   Effect of Consumption on the Coal Supply

Years the Supply Will Last with No Increase in Consumption	Percent Increase in Consumption Each Year					
	1	3	5	7	10	15
50	41	31	25	22	18	15
100	70	47	36	30	25	19
200	110	65	49	40	31	24
400	161	86	62	49	38	29
700	209	104	73	57	44	33
1000	241	116	80	63	48	35

newable resource by a constant percentage. For example, a 1000-year coal supply will last 241 years if we increase coal production by only 1% a year during this time. This same amount of coal will last only 80 years if we increase coal production by 5% a year for 80 years.

QUESTIONS

1. If the availability of energy promoted the rapid growth of cities in the latter part of the nineteenth century and into the twentieth century, where did the people come from? Why was the country able to get along without them doing what they were doing? Did the availability of energy play a part in this? Explain.

2. The population transition described in Question 1 was fairly smooth in the developed countries of the world. Why isn't a similar population transition as smooth today in many of the underdeveloped nations? Does the availability of energy play a part in these troubled regions?

3. In Appendix B, Czechoslovakia and East Germany are seen to have energy servant equivalents that are higher than those in many other parts of Europe. A comparison of the availability of consumer goods to the average citizen in these countries would not suggest that their standard of living is 50% higher than it is in the Netherlands. How might these differences be accounted for?

4. Using a basic unit of 2500 kcal per person per day and a population of 230 million people, use Figure 4-1 to calculate the total number of calories used in the United States in a year.

5. You are 25 years old and you earn $20,000 a year. Your employer has offered you a 40-year contract that has two different salary options. The first option is a raise of $4000 each year for the duration of the contract. The second option is a 5% raise based on your previous year's earnings.
   (a) Which is the better option at age 65?
   (b) How old will you be when you are earning about the same amount of money from each option?
   (c) What amount would you have to receive in the first option to equal the amount you earn in the second option at age 65?

6. What do the mathematical principles illustrated in Question 5 have to do with energy consumption?

7. If a country has a population that increases at a rate of 5% a year, what is the doubling time for the population? If the energy consumption per person in that country is increasing at a rate of 5% a year, what is the doubling time for personal energy consumption? If the country uses 100×10^{12} (100 trillion) kilojoules in a year, how much energy will this country be using 28 years from now?

8. If a country has a 200-year supply of coal using it at a constant rate, how many years will the coal last if coal consumption is increased by 7% each year until the coal is gone? What are some other problems associated with increasing the use of coal each year other than the fact that the supply will not last very long?

CHAPTER 5

Energy Use:
People and "Progress"?

So far we have considered the physical nature of energy—the inter-
actions of energy as a major component of the natural system. We have
also noted that people, at least some people, have been using nearly
100 times more than the energy required to keep their bodies functioning. This high rate of energy use suggests two questions: How long
can we continue using energy at this rate? Do we want to continue
doing so? Before we can answer these questions we need to look at
what we are using the energy for and where it comes from.

HOUSEHOLD ENERGY USE

Approximately 20% of current U.S. energy consumption is used within
our homes. Figure 5-1 subdivides this energy demand into typical
major household use categories. Space heating is the most energy consuming of the household uses, particularly for the northern part of our
country. Figure 5-2 traces the historical use of energy for heating.

Space Heating

Early Heating Methods.   Ancient civilization flourished in limited
areas where the temperatures averaged around 70°F. Early cultures
started in such places as northern Egypt, Palestine, Syria, southern

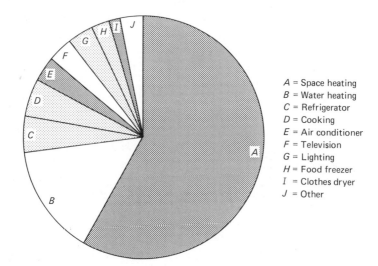

A = Space heating
B = Water heating
C = Refrigerator
D = Cooking
E = Air conditioner
F = Television
G = Lighting
H = Food freezer
I = Clothes dryer
J = Other

Figure 5-1 Typical household energy use. (Courtesy of the U.S. Department of Energy's American Museum of Science and Energy, operated by Science Applications, Inc.)

Figure 5-2 Early homes required very little heating energy (Denmark).

Arabia, and northern India. In the Americas, the Aztec, Inca, and Mayan civilizations started in the Central American highlands, an area with an average temperature of about 70°F.

As civilization progressed, people began to live in new, less hospitable areas. The Romans, although their own Italian climate was quite favorable, expanded as far north as England. To cope with the cold and damp, many of the larger, more elaborate, Roman buildings in England were constructed with a space below the floor into which hot air was piped from a "furnace" to heat the room. It is believed that this heating method, called a "hypocaust," may have been developed by the Greeks, who colonized much of what is now Turkey. The temple located at the ancient city of Ephesus was centrally heated by the burning of lignite coal.

Although this relatively sophisticated form of heating was available, most buildings, especially homes, depended on open fires and braziers. Even the larger houses in England during the twelfth century consisted mainly of a large, high-roofed hall. A blazing log fire was built in the center of this hall. The smoke simply rose, finding its way out through the roof as best it could. A separate bedroom for the master and mistress was a possibility (but a very recent development). Generally, cooking was also done over the central fire, but in the dwellings of the wealthy, the food might have been prepared in a separate building. Typically, the occupants of the houses sat or lay around this fire in the order of their relative importance, the most important being allocated the closest space.

As civilization progressed, heating methods changed, so that by the mid-fifteenth century, fireplaces with chimneys, located on the side of a room, were utilized for both heating and cooking (see Figure 5-3). This was a significant advance and at least in northern Europe, was one of a number of events that signaled the end of the Dark Ages. Central heating for homes was unheard of until relatively recent times (the twentieth century). Figure 5-4 charts the increase in heating use through the centuries. Note that as our homes have become larger and warmer, our home heating requirements have risen correspondingly.

Modern Heating Methods. Most buildings today are heated by central heating systems, where the heat for the entire building is generated at one point, a furnace. A conventional furnace is an enclosed structure built of metal, brick, or a combination of these, which can burn coal, oil, or natural gas to produce heat.

Two common heating systems are in use: warm-air heating and steam/hot-water heating systems. Radiant heating is another possibility.

Figure 5-3 Open fireplaces, located at the side of a room, were a significant advance for both heating and cooking.

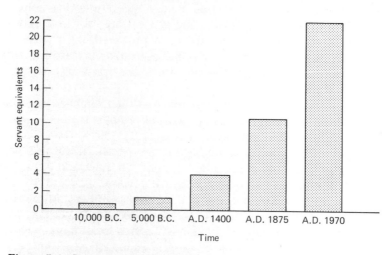

Figure 5-4 Increase in heating energy use through the centuries. (Courtesy of the U.S. Department of Energy's American Museum of Science and Energy, operated by Science Applications, Inc.)

In *warm-air systems*, air is heated in a chamber surrounding the furnace. As the air is heated, it becomes less dense and rises through large pipes, sometimes with the help of a blower, then through hot-air registers into the rooms. The cooler air returns to the furnace through a separate cold-air register and piping system. A warm-air heating system thus heats by *convection*. Convective heating occurs whenever warm room air mixes with cooler room air. The heat is spread by physical transport of the warmer substance. A clear example of convective heat transfer is the mixing of warmer and cooler water which occurs when a pan of water is heated on the burner of a stove.

Steam and hot-water systems require that radiators be used. The water is heated in a boiler above the firebox. If the quantity of heat available is sufficiently large, the water is vaporized to steam. The steam rises through pipes, then enters the radiators, where it condenses, releasing much of the heat that it absorbed when it was converted originally from water to steam. The water returns to the boiler through the same pipes as those in which it rose as steam. Hot-water systems are similar except that the quantity of heat available for release to the room is significantly less, and a separate piping system is required for the cool-water return.

The radiators used in steam or hot-water systems give off heat primarily by convection and radiation. In contrast to convective heating, radiative heating does not require the presence of any material in order for the heat energy to be transferred. The heat, as infrared waves, travels through space. When the waves strike an object, they transfer their energy totally to that object. The molecules that comprise the object vibrate more rapidly. The vibrational energy of the molecules of a substance is indicative of the temperature. The relative amounts of heat transferred by convection and radiation depend on the temperature. Steam-heated radiators transfer more heat by radiation, but the lower-temperature hot-water radiators transfer more by convection. The quantity of heat given off depends not only on the temperature but also on the shape of the radiator. Increased heat transfer can be accomplished either by raising the temperature or by increasing the surface area of the radiator.

Radiant heating generally consists of a coil or series of hot-water pipes placed in the floor. The circulating hot water radiates heat into the room, eliminating the need for radiators. Hot air could similarly be carried through ducts in a tile floor, or electric current could be used to generate radiant heat. In the latter case, the heat is generated by the resistance offered by a material to electric current flowing through it.

Each of these heating methods has advantages and disadvantages. Warm-air heating is inexpensive to install, and it can heat a house quickly, but it is less efficient than a steam or a hot-water system. Steam and hot-water systems, additionally, are more dependable, provide a more uniform distribution of heat, and by proper placement of radiators, can provide heat just at those locations where the need is the greatest, such as in front of windows or in vestibules. However, heating with radiators typically leads to higher-temperature air near the ceiling and lower-temperature air at the floor. The temperature stratification may be as much as 20 to 25°F: The ceiling temperature might be 76°F while that at the floor is only 53°F. Radiant heat does not, itself, significantly heat the air in a room, only the objects themselves; hence it is more efficient than the other systems, for it provides comfort at a lower room temperature. It is also designed to overcome the temperature stratification observed when heating with radiators.

Most central heating systems in the United States burn fuel oil, natural gas, coal, or coke. Natural gas, coal, or coke can be burned directly; fuel oil is preferably vaporized and mixed with air prior to combustion.

Fuel oil combustion can be accomplished in either of two ways. In pot-type burners, the oil flows into a shallow depression in the bottom of the furnace. In gun-type burners the fuel oil is sprayed through a nozzle under air pressure. The oil is simultaneously atomized and mixed with air. This leads to a more efficiently burning mixture. Both fuel oil and natural gas furnaces are controlled automatically by thermostats. When the temperature drops below a certain preset temperature, the thermostat makes an electrical contact which starts the oil or gas burner in the furnace. When the room temperature has increased to another set point, the thermostat again turns down the burner.

Although most homes in the United States no longer use coal or coke as fuel, it is a feasible alternative, particularly for larger building complexes. Although hand-fed systems are a possibility, most coal or coke systems use stokers to feed the fuel into the furnace. A stoker typically consists of a large screw feeder whose rotational speed can be varied by a thermostat to regulate the feed rate of fuel to the furnace, hence to regulate the heat available. This method of control is not as precise as with natural gas or fuel oil systems. Heating by electricity is clean and simple, but in most areas of the United States the cost of electric power is greater than that of other fuels.

Figure 5-5 lists six possible heating methods and their efficiencies. Heat pumps, rated as 300% efficient, might appear not to be described by the laws of thermodynamics. This efficiency is deceptive in two respects: (1) it is achievable only under optimum conditions, and (2) it considers only the energy the consumer invests, not that supplied by the external environment.

A heat pump operates like a refrigerator. It uses electrical energy to extract heat energy from the colder outside region. A cold liquid refrigerant is circulated through a coil located outside, typically buried underground. As the very cold liquid travels through the coil, it extracts heat from the ground. This heat vaporizes the liquid. The vapor then enters a compressor, where the high pressure increases the temperature of the vapor. The hot vapor is then passed through another coil, around

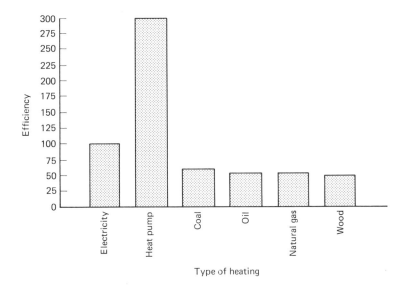

Figure 5-5 Efficiencies of various heating methods.

which an airstream flows. The hot vapor transfers much of its heat to the airstream. This warm air is circulated to the building. The vapor is then allowed to expand, and in the process it is converted back to a cold liquid. It then flows back to the ground coil, and the cycle can be repeated. The greater the temperature difference between the inside and outside spaces, the lower the efficiency. A heat pump operates very efficiently when the outside temperature is 40 to 50°F; when the temperature is very low, however, the efficiency plummets.

To retain adequate efficiency during the cold winter months, a relatively warm external source of heat is required. The heat exchange coils must be located below the frost line, where temperatures are slightly higher. Deep wells in which the water temperature may stay significantly above the outside temperature are an ideal source of heat energy. A heat pump is also a more feasible alternative in the southern United States than in the cold northern climates.

Heat pumps have the advantage that simply by reversing the direction of operation, they can serve as both furnaces and air conditioners. They can also be combined with more conventional heating or cooling methods to improve the overall efficiency of space heating.

Conventional coal, oil, or natural gas furnaces have relatively low efficiency (a maximum of about 70%, except for the new pulsating gas furnaces) because much of the heat is lost up the chimney. Good maintenance can improve significantly the efficiency obtained with any type of furnace.

Heat losses in homes.   One of the major causes of wasted heat is *infiltration:* air movement into and out of a house. Tiny cracks around windows, doors, siding, and along the foundation let cold air in and warm air out. It is estimated that infiltration losses usually account for about 35% of the heat loss in a typical home. This will decrease as insulation use and building standards rise.

Infiltration losses are fairly easy to minimize. Weather stripping and caulk can be applied in cracks around doors, windows, air conditioners, and exterior trim. The cost is generally low (less than $100), and the savings realized by lower heating and cooling costs should more than cover this investment in less than a year. Both caulk and weather stripping act in the same way: They prevent the seepage of cold air in (and hot out), decrease drafts, and keep out dust and dirt. Caulk will also prevent the seepage of water into the home, where it could cause rot and paint peeling.

Caulking should be done at all stationary joints. Caulk comes in a number of varieties. Oil-based caulk is the oldest type. It is inexpensive, but has a life expectancy of 2 years at the most. It will dry out, shrink, crack, fall out, and have to be replaced. Better caulking is now made of urethane, silicone rubber, or an acrylic latex. Some of these varieties have up to a 20-year guarantee.

Many modern doors and windows have weather stripping included at the factory which probably will not need replacement for years, if at all. Older homes are more likely to need updating in this area. Weather stripping should be applied to all movable joints, such as windows and doors. Common weather-stripping materials include spring metal, rolled vinyl, and adhesive-backed foam. The specific type preferred will depend on the type of window or door.

Another possible type of infiltration loss is within fireplaces. Contrary to what one might expect, burning wood or similar fuel in a typical home fireplace during the cold winter months will usually increase heating costs. In a traditional fireplace, the burning fire consumes large quantities of air for combustion and drafting. If a fireplace burns during the evening hours and then goes out while the residents are sleeping, the open chimney flue provides an exit for less dense warm air from inside the house. This is like having a window open. The furnace therefore has to be on longer to heat the house. Glass fireplace doors can help reduce this heat loss significantly.

The amount of heat that is actually lost through the walls, windows, and roof is dependent on their insulating values (see Figure 5-6), their areas, and the difference between inside and outside temperatures. Most of the heat is lost through *conduction*, a slow process by which heat is transmitted by contact of a colder surface with a warmer one. In conduction the heat is transferred through a substance by the

(a)

(b)

Figure 5-6 The amount of heat lost from a home through the walls, windows, and roof is dependent on their insulating values: (a) the picture of a house taken under normal photographic conditions indicates a well-built structure; (b) the same house, photographed using infrared (heat)-sensitive film. Note the lighter areas where heat escapes. (Courtesy of the U.S. Department of Energy's American Museum of Science and Energy, operated by Science Applications, Inc.)

increased motion of the molecules of the conducting substance. The temperature of a solid substance is determined by how fast its molecules are vibrating. In conduction, rapidly vibrating molecules strike their neighbors, causing them to move faster. This transfer of motion thus transfers the heat energy through the object.

Heat always flows by conduction whenever there are two substances that are in physical contact and at different temperatures. Some materials are better heat conductors than other materials. Metals, in general, are good thermal conductors; nonmetals, such as glass, wood, and paper, tend to be thermal insulators, or poor conductors.

In building products and in construction, the heat flow by conduction is reduced by introducing an insulating material. For example, the air space between the inside and outside windows is a good insulator because it is much more difficult to transfer heat energy from one gas molecule to another than it is to transfer the energy from one molecule to another in solids. Commercial insulation, such as that used within walls, is made of a lightweight material with a multitude of fine air spaces within its structure. The insulating ability of various materials is expressed as its R *value*. The units are typically hr-ft^2-$^\circ$F/Btu. Better insulating substances have higher R values. Some typical R values for different common types of insulation are included in Figure 5-7.

In addition to discussing the R value, which measures resistance to the flow of heat, one can discuss the U value, which is a measure of how fast heat will pass through a given section of the construction. These two values are the reciprocal of one another ($R = 1/U$).

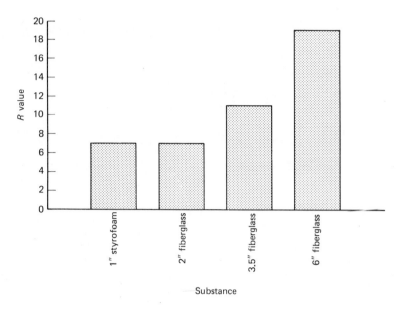

Figure 5-7   Typical R values for common insulating materials.

From information on the R values of the various structural components, it is possible to calculate the heating requirements for a building. The rate of heat loss, Q, can always be calculated by

heat loss (Btu/hr)
= area (ft^2) × U (Btu/hr-ft^2-°F) × temperature difference (°F)

For example, consider a picture window 8 ft wide by 5 ft high. Its area is 40 square feet. Assume that the outside temperature is -20°F and the inside temperature is 70°F. If the window is a single thickness of glass, $R = 0.88$ hr-ft^2-°F/Btu. Since $U = 1/R$, $U = 1/0.88 = 1.14$ Btu/hr-ft^2-°F. The rate of heat loss is then

$$Q = 40 \times 1.44 \times [70° - (-20°)] = 4068 \text{ Btu/hr}$$

If a substance (a wall, for example) is comprised of several layers of different materials, the R values of each layer can be added to get the total; for example,

$$R_{3\frac{1}{2} \text{ in. insulation}} + R_{25/32 \text{ in. sheathing}} + R_{3/4 \text{ in. siding}}$$
$$+ R_{\frac{1}{2} \text{ in. gypsum board}} = R_{\text{total}}$$

The conduction heat loss for the total building is the sum of the heat losses for the individual components. The rate of the heat loss for a house thus would be

$$Q_{\text{conduction}} = Q_{\text{walls}} + Q_{\text{ceiling}} + Q_{\text{floor}} + Q_{\text{glass}} \text{ (doors and windows)}$$

To simplify the calculations, all components with an identical R or U value can be lumped together and the total area used in the calculation.

The effect of infiltration can also be included in the overall calculation:

$$Q_{\text{infiltration}} = \text{volume heated space (ft}^3) \times 0.018 \times \text{number of air changes per hour} \times \text{temperature difference (}^\circ\text{F)}$$

An estimate of one air change per hour is frequently used. The overall rate of heat loss for a building is the sum of $Q_{\text{conduction}}$ and $Q_{\text{infiltration}}$.

Appliances

Electric household appliances also consume large quantities of energy. The average wattage and estimated kilowatt-hours of energy consumed annually by a variety of appliances are listed in Appendix C.

The home hot-water heater is invariably the largest single energy-consuming appliance. The energy required for a hot-water heater depends on the type of heater. For a typical household of two to four people, gas hot-water heaters consume 7% of the total energy, while electric water heaters may consume 12%. Hot-water production is the third largest item of energy use in most households, following home heating and transportation. Overall hot-water production requires 4% of our national energy budget. The major household uses of hot water, in decreasing order of consumption, are bathing, laundry, dishwashing, and cooking.

TRANSPORTATION ENERGY USE

About 25% of U.S. energy demand is now dedicated to transportation. The historical trends in energy used by transportation are illustrated in Figure 5-8.

Until approximately 100 years ago, all transportation was on foot or used animal power. The average speed that people traveled was typically less than 20 miles per hour, even though the distances were sometimes great. Twenty-five hundred years ago Alexander the Great and his army marched from Egypt into India, a distance of several thousand miles.

By the mid-1850s, the railroads, powered by the steam locomotive, became a very important means of transportation in the United States (see Figure 5-9). At that time, the country was entering its greatest era of expansion. The railroads permitted the easy shipping of western

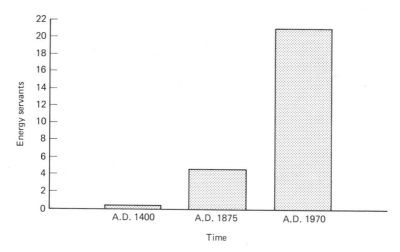

Figure 5-8 Historical trends in transportation energy use. (Courtesy of the U.S. Department of Energy's American Musuem of Science and Energy, operated by Science Applications, Inc.)

Figure 5-9 The early steam locomotive was a major factor in the expansion of the West. (Courtesy of the U.S. Department of Energy's American Museum of Science and Energy, operated by Science Applications, Inc.)

grain and cattle to eastern markets, thus encouraging settlers to purchase land in more distant and remote areas of the country. This led to the eventual growth of new markets, then large cities, then the West itself.

Although the steam locomotive increased man's maximum traveling speed by two or three times, it did not provide people with the same capability that the invention of the automobile did (see Figure 5-10). The automobile provides not only a means of transportation, but often becomes an extension of the person. It can provide a portable home and it can also greatly increase one's "psychic room": The ability to move over large areas quickly allows one to use these areas to provide a physical and psychosocial environment upon which one can draw. However, the speed of travel has not changed all that much. The maximum speeds have increased by a factor of 3 or 4 if we compare the horse or bicycle to a car. In the past, each year people could travel several hundred miles; now they may travel several thousand miles, an increase by a factor of 10.

Increased use of the internal combustion engine, such as in the automobile, has had two major effects on the environment: pollution

Figure 5-10 Automobiles are used for a major share of the passenger-miles traveled in the United States. (Courtesy of the U.S. Department of Energy's American Museum of Science and Energy, operated by Science Applications, Inc.)

and increased energy use. The automobile is the primary culprit in the formation of smog and carbon monoxide (see Figure 5-11). Although this is not a major problem in some areas, it frequently is in cities such as New York and Los Angeles.

Pollution and energy consumption are directly related in subtle, perhaps unpredictable ways. For example, gasoline "shortages" have caused many California drivers to use leaded fuels in cars designed only for nonleaded gasoline. As a result the catalytic converters used for pollution control have been destroyed, thus causing the pollution levels to rise drastically again.

Transportation of both passengers and freight can be very energy consuming. The relative efficiencies of the various types of transportation, both within a city (urban) and between cities (intercity and/or rural), are indicated in Figure 5-12. Automobiles and airplanes together account for 98% of our total mileage, and these two are by far the least energy efficient means of transportation. There are so many automobiles in the United States that it would be possible to put every living American in the front seat of an American automobile, two people per car, with no one in a back seat.

An automobile is typically only 5% energy efficient (Figure 5-13); that is, only 5% of the total energy available in crude petroleum is used to move the car. The remainder of the energy is lost, primarily as waste heat. The average value in 1980 was less than 15 miles per gallon (mpg). The government set a 27.5-mpg standard to be met in the future with

Figure 5-11 The automobile is the primary culprit in the formation of smog in areas such as Manhattan. (Courtesy of the U.S. Department of Energy's American Museum of Science and Energy, operated by Science Applications, Inc.)

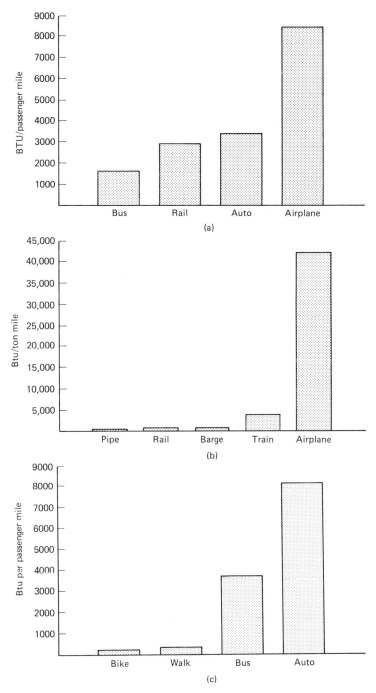

Figure 5-12 Transportation energy efficiency: (a) intercity passenger transport; (b) intercity freight transport; (c) urban passenger transport. (Courtesy of the U.S. Department of Energy's American Museum of Science and Energy, operated by Science Applications, Inc.)

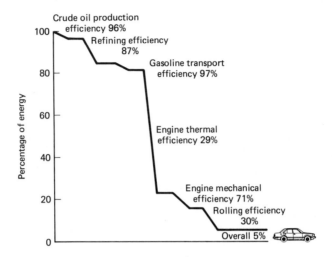

Figure 5-13 Automobile energy efficiency. (Courtesy of the U.S. Department of Energy's American Museum of Science and Energy, operated by Science Applications, Inc.)

hopes of 50 mpg by the year 2000. To avoid fines for not meeting these mandates, auto companies are decreasing the size and weights of their car models. The size and weight of a vehicle are directly related to energy consumption. Decreasing the weight by one-half nearly doubles the gas mileage. Data also show that substantial savings in gasoline consumption could be accomplished by reducing air resistance, such as using shields on truck cabs; by driving on better, optimally inflated tires; by not exceeding the 55-mph speed limit; and by trying to design better transmissions and more efficient engines.

This is a significant area in which to reduce energy consumption because approximately one-half of our crude petroleum is imported, primarily from the Middle East, and one-half of that is converted to gasoline. The importation of oil has many implications, which are discussed in Chapter 7.

The convenience of the private car and of commercial and private aviation has greatly increased our potential for business and pleasure. The jet age has made possible previously unheard of activities. A person can commute to meetings from Chicago to Boston, to Philadelphia, to Atlanta; he or she can get in a half-day's work in both Paris and Washington. Airplanes (Figure 5-14) are now used for 9.3% of travel within the United States.

In recent years we have seen a demise of public transportation such as buses and railroads (see Figure 5-15). Only a slow recovery, over a period of about 25 years, is possible or likely.

In the United States, 84% of all intercity traffic is by auto and less than 1% is by rail. This is in contrast with the other industrialized nations of the world, which have been racing toward development of completely new, high-speed rail systems. Although one attraction of these trains is their speed, their real success lies in that they are very efficient and run on electric power. The cost of operation and maintenance of efficient electrically powered rail systems is only 70 to 75% that of a comparable diesel-powered system.

Japan has had electrically powered "bullet trains" running between Tokyo and Hakata since the mid-1960s. Over 2 billion passengers have utilized these trains, which have been very profitable and extremely safe and reliable. The Japanese started services to the north of Tokyo with second-generation high-speed trains in 1982.

France now has a fleet of 160-mph electric trains considered superior to the automobile and airplane in terms of cost, speed, safety, and comfort. Two months after service between Paris and Lyons was started in September 1981, 15,000 passengers were using the service daily. If the energy required for construction is not factored in, the per passenger energy use is equivalent to that of a motorbike trip.

Why hasn't the United States joined this race? The primary reason is the physical design of the track system. Most track in the United

Figure 5-14 Airplanes, a very energy inefficient means of transport, are now used for about 10% of travel in the United States. (Courtesy of the U.S. Department of Energy's American Museum of Science and Energy, operated by Science Applications, Inc.)

Figure 5-15 Public transportation (here in Glasgow, Scotland) is very common in Europe and other areas, but not in the United States.

States is flat, designed to accommodate long, heavy freight trains which travel, on the average, about 20 mph. This design is totally incompatible with the new high-speed trains. Trains traveling 55 mph or greater require superelevated tracks, canted in the curves (that is, rails with a sloping surface), with the outer rail higher than the inside rail. If these tracks were built, they would not serve the slower trains. At slow speeds, most of the weight would be on the inner rail, which would suffer excessive wear. The heavy cars might even tip over. A totally independent track system is thus required. Its construction will be very expensive in terms of both monetary and energy investment.

High-speed systems are planned, however, for the United States in the near future. The Ohio Rail Transportation Authority is constructing a high-speed, all-electric passenger rail system along these three routes: Cleveland–Columbus–Dayton–Cincinnati, Toledo–Cleveland–Akron–Youngstown, and Detroit–Toledo–Columbus. The American High-Speed Rail Corp., based in Washington, D.C., is planning a system between Los Angeles and San Diego. It is expected that 25% of the estimated $2 billion investment will come from Japanese investment groups.

COMMERCIAL ENERGY USE

Our stores, schools, hospitals, and government buildings constitute over 14% of the total U.S. energy demand. Much of this energy is used for heating, air conditioning, refrigeration, and lighting. Large amounts of energy are also required for packaging the products sold by commercial businesses. For example, a few years ago the paper and plastic containers used to package McDonald's food required enough energy to provide the electric power needs of Pittsburgh, Boston, Washington, and San Francisco.

Disposing of solid waste also requires large quantities of energy. For example, every day the average American throws away one beverage can or bottle, each of which required about $\frac{1}{2}$ kilowatt-hour of energy to produce. This is sufficient to light five 100-W bulbs for one hour and thus could meet the daily lighting needs of the average U.S. family. Stated another way, the energy required to produce each disposable beverage container is equivalent to filling it one-third full of gasoline before throwing it away.[1] The typical recycled bottle could, on the other hand, be reused 15 times. Aluminum can be recycled using only 5% of the process energy required to manufacture it from the original bauxite ore. The total U.S. energy requirement for solid waste

---

[1] "The Second Law," Northwest College and University Association for Science, Richland, Wash., May 1977.

disposal is equivalent to the total output of two 1000-megawatt power plants.

This is, however, not the complete story. Recycling of anything, such as cans, requires that a low-density distribution be gathered together into a high-density distribution. Any such density change is very expensive in energy, time, effort, and money. If you include the energy and money needed to gather together beer cans, it is very doubtful if such recycling is energy or money efficient.

Except for a few hundred pounds in space on satellites, there is as much copper, for example, on earth now as there was a million years ago. But the use of copper has led to the replacement of relatively concentrated deposits of the ore by widely dispersed copper in machines and refuse dumps. Enormous amounts of energy are required to reconcentrate the copper in dumps back to a usable form.

This problem is the same as arises with the "exhaustion" of our fossil fuels. When the coal deposits are nearly exhausted, there will still be nearly as much carbon on the earth as ever, but it will be as CO_2 and other dispersed forms. These forms are unsuitable for use as a fuel.

INDUSTRIAL ENERGY USE

Modern industry first started making advances in England at the beginning of the Industrial Revolution, around 1750. Prior to that time, there was little division of labor. Most farmers were essentially self-sufficient. In the cities, some manufacturing was carried on by craftsmen producing such things as hardware, cloth, and weapons. There was really nothing, however, that could be classified as true industry in the modern sense. Around 1750, there was a fortunate combination of two unrelated factors which led to the development of modern industry. First a few small merchants and traders became wealthy, primarily due to trading with the colonies. Simultaneously, a number of scientific investigations led to important technological inventions. It was the combination of these two factors, money and technology, that led to modern industrial development.

In spite of the Industrial Revolution, industrial energy consumption remained fairly modest until recent years. Much work was still done using hand or animal power (see Figure 5-16). In recent years, however, the high price of labor in the United States combined with the availability of cheap energy has led to the high rates of energy consumption that we are now experiencing. The historical trends of combined industrial and agricultural energy consumption are summarized in Figure 5-17.

In the United States, industry consumes about 41% of the U.S.

Figure 5-16 An early craftsman.

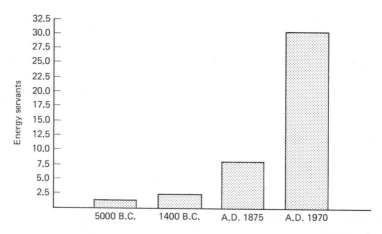

Figure 5-17 Historical trends in combined industrial and agricultural energy consumption.

total energy budget (see Figure 5-18). The top 10 energy-consuming industries are the following, in decreasing order:

1. Chemicals and allied products
2. Primary metals
3. Petroleum and coal products
4. Stone, clay, and glass products
5. Paper and allied products
6. Food products
7. Fabricated metals
8. Transportation equipment
9. Machinery (except electrical)
10. Textile mill products

Table 5-1 provides a breakdown for the sources of energy and the total Btu used by the largest industries within one particular area in the United States. Although these data are specific for Brown County, Wisconsin, these combinations of energy sources can be considered typical for many areas in the industrialized world. The total energy

(a)

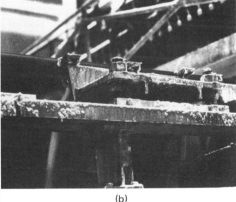

(b)

Figure 5-18 Modern industry consumes large quantities of energy: (a) aluminum production comprises 10% of the total U.S. industrial energy consumption. (b) Paper production involves large quantities of energy in the drying steps. (Courtesy of the Wisconsin Paper Council.)

TABLE 5-1 Industrial Energy Use (Btu $\times$ 10^9), Brown County, Wisconsin

Industry	Natural Gas	Electricity	Fuel Oil	Coal	Total
Paper	9,690	3,880	21,300	10,800	45,700
Food processing	888	136	320	76.3	4,300
Chemicals	108	8.53	525	41.5	683
Metals	178	42.2	252	66.5	539
Machinery	141	39.5	207	11.1	399
Wood	94.1	20.2	31.3	3.97	150
Construction	33.9	6.68	43.9	—	84.0
Electrical	26.7	11.5	16.8	—	55.0
Petroleum	4.94	1.17	42.3	4.16	52.6
Quarries	0.611	2.49	0.087	0.0964	3.29
Total	11,165	4,148	22,694	11,003	51,966

Source: Robert Lanz.

used by each industrial category has two parts: (1) the energy intensiveness of the industry, that is, the energy input required to produce a given amount of product; and (2) the concentration of that type of industry in the given area. The relative amounts from each source of energy, however, are fairly constant regardless of the geographical area.

In many industries there are significant energy savings possible by process changes. For many years energy in the United States was inexpensive (and even getting cheaper) and labor was expensive. In much of Europe the opposite has been true. The Europeans have, therefore, developed industrial processes that require more labor but less energy, whereas we have done the opposite (see Figure 5-19). For example, 60% of the cement plants in the United States use a "wet process" which requires about twice as much energy as a "dry process" used almost exclusively in Europe. Although the capital investment is usually too great for U.S. industry to change processes *only* to conserve energy, new plants and updated plants can—and are—utilizing the less-energy-

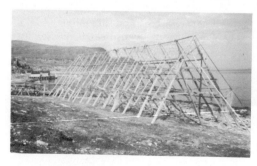

Figure 5-19 Different life-styles require less energy. The Norwegian fishing industry, for example, depends on solar energy for fish drying, on racks similar to these.

intensive processes. Developmental work is also under way in the aluminum industry, where the aluminum purification process requires over 10% of the total U.S. industrial energy demand. This is also true for the steel industries. The reason the chemical industry is ranked so high is because it uses energy sources (petroleum) as its raw materials as well as in the processing. Studies are now under way to find suitable substitute raw materials for the production of many synthetic fabrics, plastics, rubber, fertilizers, and pharmaceuticals.

ENERGY IN AGRICULTURE

Early hunting and gathering societies usually consisted of relatively small groups of people and, unless food was abundant, much of the energy of the group was needed to obtain and process food. This left little time for other activities. When some animals were domesticated, the total amount of energy used for food production increased because now the animals had to be maintained, but in most cases the animals did not compete with people for food because the animals ate cellulose material, such as grasses, which people are unable to digest.

There are several advantages for people in including domestic animals in their agricultural practices. One advantage is that domestic animals such as oxen and horses have the strength to do many tasks that one person could never do alone, such as moving heavy objects, pulling plows that reach several inches into the soil, and carrying heavy loads. The domestic animal could also serve as a source of food and fertilizer. Perhaps the most significant benefit for the individual and for society was that by using domestic animals people reduce the number of hours required to produce a crop from the land to about one-half the number of hours required when all the work was done with human labor.

The usefulness of human labor was greatly increased when farm machinery became available that was powered by gasoline. We complain about paying in excess of $1 a gallon for gasoline, but from a strictly energy standpoint (calories or Btu), this is a real bargain because 1 gallon of gasoline is equivalent to about 100 hours of human work. Using a minimum wage of $3.50 per hour, a gallon of gasoline at $1.30 has an energy market value of $350 in equivalent human effort. In addition, human efforts have physical limits. For example, it would be impossible for human beings alone to provide a transportation system that travels 500 miles per hour. The change in the source of energy input in American agriculture is illustrated in Figure 5-20.

As a result of the shift from human energy to fossil fuel energy, two important changes have occurred. First, productivity of the land in

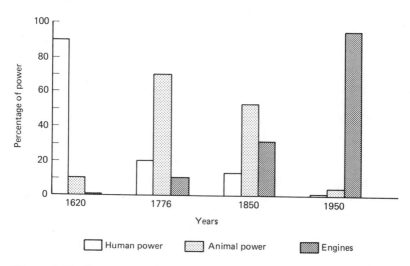

Figure 5-20 Percentage of power provided by human power, animal power, and engines. [Courtesy of D. Pimentel and M. Pimentel, *Food, Energy, and Society*, Resource and Environmental Sciences series (London: Edward Arnold (Publishers) Ltd., 1979). Based on data from E. Cook, *Man, Energy, and Society* (San Francisco: W. H. Freeman, 1976).]

the United States has increased because of the extensive use of fertilizer. In 1910, with no commercial fertilizer, corn production averaged about 30 bushels per acre. In 1980, with the addition of 250 lb of commercial fertilizer to an acre, corn production increased to about 100 bushels per acre. The number of labor hours required to produce an acre of corn also decreased dramatically. If an acre of corn is tilled entirely with manual labor, about 450 hours are required. If 1 acre of corn is tilled using modern farm machinery, the job can be done in about 5 hours. The hand-tilled acre, however, requires a total energy input about one-fourth of that required where machinery is used for tilling.

Extensive use of fossil fuels in American agriculture has decreased the human time input to about 1% of that required to do all the work 200 years ago. The total energy input, however, has increased by 400%, and nearly all of this energy comes from nonrenewable fossil fuel energy sources. This has significant implications when one considers existing world food shortages. Farmers represent less than 5% of the work force in America and they produce enough food to meet the needs of the U.S. population plus an additional 70 million people. Unfortunately, the success model of the American farmer is not readily transferable to the destitute areas of the world. Many nations that are having problems raising enough food to feed their people do not have the fossil fuel resources necessary to support the agricultural technology needed to produce this increased productivity. They may already be using 60 to 80% of their available fossil fuel energy to produce food. They also do not have the technological support system necessary to keep their machines functioning.

Many of the farms on which people are attempting to support themselves in undeveloped nations would be abandoned by U.S. standards because the soil quality is so poor, rainfall is not appropriate, or there is no way to market or store the food produced. Also, the ability to reduce labor hours in agricultural production is not a particularly attractive goal in underdeveloped countries because they are plagued with high unemployment levels.

Food Processing

The amount of energy required to transform food from its condition as a live plant or animal to a product that can be consumed by human beings varies dramatically from one food to another, and has changed in the past century. One obvious energy requirement is the necessity to cook many foods. Cooking kills germs and in many cases makes food more digestible, in addition to enhancing the flavor. In addition to cooking, energy is required for milling grains, butchering animals, and packaging and preparing foods for storage. Transportation energy costs are also significant, particularly for foods that are grown in restricted climates. Table 5-2 gives two examples of the use of energy in food preparation and distribution.

Food packaging alone accounts for enormous differences in energy. Wooden berry boxes can be produced using 69 kcal of energy, steel cans require more than 1000 kcal, and glass bottles can range from 2500 to 4500 kcal. The TV dinner tray requires nearly twice as much fossil fuel energy to produce as the food energy available from the meal it contains.

TABLE 5-2   Energy Inputs in Food (kcal)

	1 lb of Sweet Corn	1 lb of Beef
Production	450	94,250
Processing	262	179
Packaging	1,006	168
Transportation	158	163
Distribution	340	341
Total	2,216	95,096

Source: Based on data from D. Pimentel and M. Pimentel, *Food, Energy, and Society* (London: Edward Arnold (Publishers) Ltd., 1979).

Transportation costs are an important factor in determining the energy costs of foods. When food is transported across the country, 50 or more calories in energy for transportation may be required for every food calorie delivered. Food that is delivered by waterway can be transported 80 miles for the same amount of energy needed to send it 1 mile by air. Railroads can transport the food 50 miles using the same amount of energy.

The food on the plates of most Americans has traveled an average distance of about 400 miles. Because of transportation costs it is more efficient in terms of energy to consume frozen orange juice than to consume fresh oranges.

Food preparation in the home in the United States requires an average of 9000 kcal per day for each person. Differences obviously depend on what food is being prepared. In any case, Americans utilize three times the fossil fuel energy to prepare the food at home as the caloric value of the food on the plate. The type of stove used is also a significant feature in determining the efficiency of home cooking. Electric stoves are the most efficient in converting the energy supplied to the stove to heat, but the efficiency of producing electricity is only about 30% from the primary fuel. Thus natural gas is the most efficient source of primary fuel for cooking at about 33% efficiency, with electricity and wood in the range 20 to 25%.[2]

QUESTIONS

1. Consider a clapboard house described below. The wind outside registers 15 mph.

    First floor: 20 ft × 44 ft, 7 windows, ordinary window glass, 1 sliding door, 6 ft × 7 ft, thermopane. Second floor: 22 ft × 44 ft, 8 windows, ordinary window glass. There is a complete but unheated basement.

Specifications	R Values
Walls:	
Clapboard, ½ in. × 8 in.	0.85
Building paper	0.06
Wood sheathing, $^{25}/_{32}$ in.	0.98
Air space due to 2 × 4's	0.97
Gypsum lathe	0.32
½-in. plaster	0.09
Insulation (per inch)	3.17
Outside surface with 15-mph wind	0.17
Inside surface	0.68

[2] David Pimentel and Marcia Pimentel, *Food, Energy, and Society* (London: Edward Arnold (Publishers) Ltd., 1979).

The Btu/hr-ft^2-$°$F difference for ordinary window glass is 1.13 and for thermo-pane window glass is 0.57. Assume that all downstairs windows are 3 ft $\times$ 5 ft and all upstairs windows are 3 ft $\times$ 4 ft. The ceilings are 8 ft both up and downstairs.

From these data, calculate the Btu required per hour to heat this house under the following conditions:

T(outside) $= 10°$F

T(inside) $= 70°$F

4 in. of insulation in all walls and ceilings

What assumption have you made for this calculation? Would you expect the actual heating cost to be less or more than this? Why? How could you improve the heating efficiency of this house without making any major structural revisions?

2. Discuss the "energy economics" of recycling aluminum cans. Aluminum can be recycled using only 5% of the process energy required to manufacture it from the original bauxite ore. What other energy inputs are there into the entire recycling process? Who provides these energy inputs?

3. Why is it difficult for any industry to reduce its energy consumption significantly in the short term?

4. Discuss the factors that would make it difficult for the United States to make a rapid transition to increased use of mass transit even if it were deemed very feasible to do so because of increased gasoline prices.

5. Make an inventory of the electrical appliances in your household. Which of these are large energy consumers? Which consume only relatively small amounts of energy?

6. Why is it possible to answer the question "Is the American farmer the most efficient farmer in the world?" with a response of "yes" or "no" and be correct in either case?

7. What are some ways in which energy transportation costs for food could be reduced?

8. Why isn't the food required to maintain a domestic animal often not considered as competing with the food needs of people?

9. Give several reasons why the highly successful American system of agriculture cannot easily be copied by farmers in countries where there are food shortages.

CHAPTER 6

Coal: The Old Standby

There are many available sources of energy. Some of these, such as coal, natural gas, oil, nuclear power, and hydropower, are classified as *conventional* energy sources. These conventional sources are the primary energy sources for the United States at the present time. Other sources, such as solar, wind, biomass, solid waste combustion, and geothermal, are considered *alternative* energy sources. The alternative sources are those which show promise for the future.

The major conventional sources are called *fossil fuels*. The fossil fuels include coal, oil (petroleum), and natural gas. They are nonrenewable (within any reasonable time frame) and once they are gone they cannot be replaced. Currently, they are our major fuels for electric power production, for our automobiles, and for space heating the majority of our homes. This chapter will consider the primary fossil fuel found in the United States, coal (Figure 6-1).

FORMATION

The formation of fossil fuels requires a long period of time because it involves an intricate set of geologic processes. There can be some variation in these processes from one location to another, and from one fuel type to another, but in general, the formation of all the fossil fuels is quite similar.

Figure 6-1 Coal is the primary fossil fuel found in the United States. Per Btu, it is much less expensive than oil or natural gas. (Courtesy of the Wisconsin Department of Natural Resources.)

Coal originated from plant material that grew in upland bogs, in coastal or near-coastal swamps, or in delta plains. A similar existing location might be the Florida Everglades. The size of coal formations ranges from a few acres to hundreds of square miles. The large coal fields were originally flooded by either an increase in sea level or subsidence (sinking) of the land. Burial of the plant material resulted in the absence of oxygen, so the material could not be broken down into the normal decay products, carbon dioxide and water. The weight of the material covering the organic matter produced pressure, which in turn produced heat. The heat and pressure produced chemical and physical changes which resulted in the formation of fossil fuels.

Coal is produced by gradual transformation through several stages from peat to subbituminous to anthracite coal. The percentage of carbon increases in this process because other elements, such as hydrogen and oxygen, are removed. Generally, the more advanced stages of coal are better fuels because there is more heat energy available from each pound of coal and there are also smaller amounts of air pollutants produced when the coal is burned.

Some of the major charactcristics of the various types of coal are summarized in Table 6-1. Anthracite and bituminous coal can have the

TABLE 6-1 Major Characteristics of Coal

Class	Percent Fixed Carbon	Percent Ash	Percent Volatile Matter	Percent Sulfur	Calorific Value (Btu/lb), moist
Anthracite	92–98	2–4	2–8	0.7–1	12,900–14,000
Bituminous	69–78	$4\frac{1}{2}$–15	22–31	0.7–4	10,000–14,000
Subbituminous	—	7–15	—	0.5–0.9	7,500–10,500
Lignite	—	5–10	—	0.5–0.7	6,300–8,300

same calorific value yet different percentages of fixed carbon because bituminous coal also contains 5 to 6% hydrogen, which contributes to the available energy. The calorific value is determined primarily by the percentages of moisture and ash.

COAL RESOURCES AND RESERVES

As with all fossil fuels, the actual quantity of the coal that can be economically made available is the subject of considerable debate. There are several reasons for this:

1. What does "available" mean? Many deposits may be in nonaccessible regions—located under towns, parks, and other regions where it is not feasible to extract them.
2. For some deposits it may be too costly, either in direct dollar expenses or in terms of the amount of energy required, to mine the coal. It is not feasible to extract an energy source if more energy is required in the process than can be realized from the use of that fuel.
3. It is very difficult, without extensive testing and mining, to estimate how much coal is actually in a deposit.
4. It is also very difficult to identify where the deposits are, especially those lying at depths over 1200 meters, although it is likely that there are considerable resources at these lower levels.

Figure 6-2, based on estimates from the 1977 World Energy Conference, illustrates graphically the difference between the geological resources (the resources that *may* someday become of economic value) and the technically and economically recoverable reserves which are feasible under the conditions now prevailing. It is easy to see that there is not a direct correlation in each instance between the total resources available and the reserves that are now feasible for recovery. In this table, hard coal includes anthracite and bituminous coal; brown coal includes subbituminous coal and lignite. The energy value that separates these categories is about 24,000 J/kg (10,500 Btu/lb), the hard coal having the higher values.

At the 1977 World Energy Conference, global coal resources, as ultimate quantities, were estimated to be 10.1 trillion tons. This is enough to last 1200 years at the then-present world consumption rates. The global coal reserves, those which were well identified and could probably be produced profitably, were only 837 billion tons, about 5% of the total estimate. Even if, as is likely, this estimate of useful reserves could double in the next 40 years, there is still a considerable disparity between these two figures.

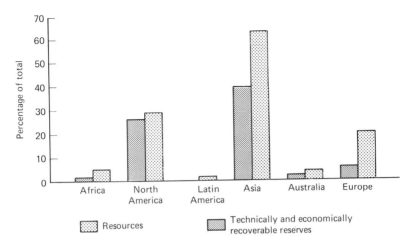

Figure 6-2 World coal resources and reserves. (Courtesy of the U.S. Department of Energy's American Museum of Science and Energy, operated by Science Applications, Inc.)

LOCATION OF THE DEPOSITS

Figure 6-3 illustrates the locations of the major coal supplies in the world. The USSR, although it has tremendous resources of coal, has difficulty mining the coal because many of the deposits are located in quite inaccessible regions of Siberia. The cost to mine these deposits is currently prohibitive.

Most of Western Europe has very little coal, in fact, very little of any of the fossil fuels. This has led to the development of energy-saving

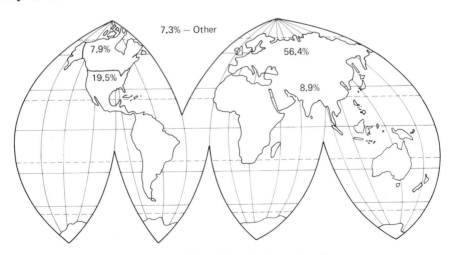

Figure 6-3 World distribution of coal.

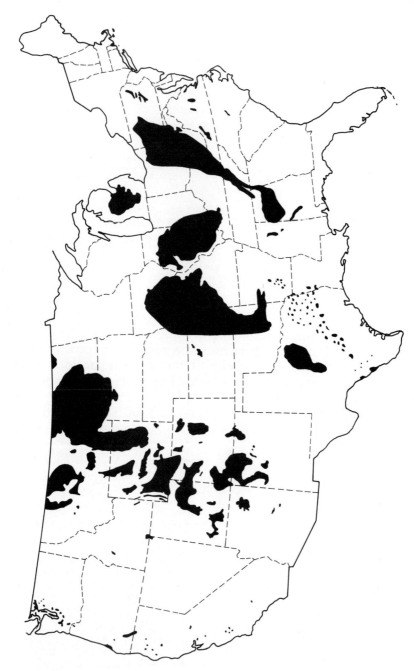

Figure 6-4 Distribution of coal in the United States. (Courtesy of the U.S. Department of Energy's American Museum of Science and Energy, operated by Science Applications, Inc.)

technologies ranging from automobiles to manufacturing plants. As discussed in Chapter 5, in Europe for many years labor has been relatively cheap but energy has been expensive, whereas in the United States the reverse has been true. As a result, Europeans and Americans have developed quite different life-styles and technologies.

Many of the developing areas of the world are without significant coal supplies. Since energy is a critical ingredient for any type of industrial development, this factor has increased their dependence on other (industrialized) countries and inevitably hampers their further growth.

The United States is blessed with relatively large deposits of coal. The distribution of this coal is shown in Figure 6-4. There are significant differences between the coal fields in the eastern and western United States. The eastern deposits are often deeply buried, hence they are mined by underground shaft mines, whereas those in the West are generally located closer to the surface and thus are strip mined. Eastern coal also contains more sulfur impurities, resulting in more pollution problems when the coal is burned. Western coal usually has a lower heat value per pound, so more coal must be consumed to obtain an equal amount of energy.

USE PATTERNS

Coal was the first of the fossil fuels to be used. Figure 6-5 illustrates the historical patterns for the use of the various fossil fuels. By approximately 1880, half of our energy was supplied by coal (the remainder being wood). In terms of total energy demand, coal use peaked during the 1920s.

In the United States it is expected that coal use will continue to increase for the near future, because (1) it is the only fossil fuel that we have in abundance, and (2) the technology for its use is well developed. The estimated reserves of coal within our own borders will probably last less than 100 years, however, if we keep increasing our energy demands at the same rate as in the late 1970s. However, stabilizing or even slightly lowering our energy demand can extend the lifetime of these reserves for several centuries.

Figure 6-6 compares estimates of probable rates of future world coal use from two sources: the Coal Industry Advisory Board (CIAB), which functions under the auspices of the Organization for Economic Cooperation and Development, and the World Coal Study (WOCOL). A more detailed analysis is presented in Appendix H.

As can be seen, these estimates vary significantly, particularly for heat and electricity generation. Coal conversion will probably play an increased role near the end of the century, but it will not be a major factor in the projected tripling of coal use by the year 2000.

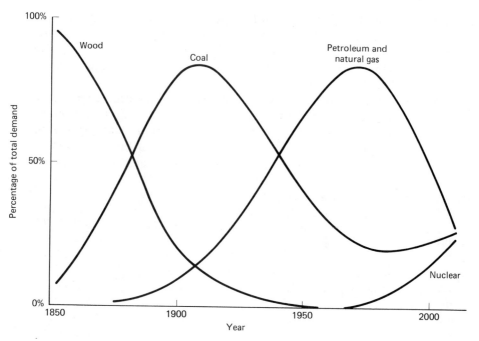

Figure 6-5   U.S. sources of energy from a historical perspective. (Courtesy of the U.S. Department of Energy's American Museum of Science and Energy, operated by Science Applications, Inc.)

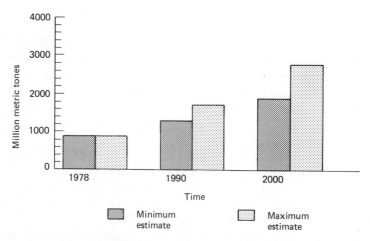

Figure 6-6   Projected coal use, 1990 and 2000.

ENVIRONMENTAL EFFECTS

There are two difficulties associated with greatly increasing the U.S. use of coal in the near future.

1. Extraction of the coal, whether by underground or strip mines, leads to environmental damage. This is particularly true for the very desirable western coal, primarily due to the geography of these sites.
2. The burning of coal leads to air pollution.

Mining

Strip Mines. The two types of coal mining, strip mining and underground mining, are quite different from one another. Figure 6-7 illustrates a typical strip (surface) mine. The overburden (soils, rocks, etc.) removed from a rectangular area is placed alongside the immediate mining area. The coal is then removed by power shovel, front-end loader, or other similar device. The process then moves to a new area. The overburden from the new area is placed into the area just excavated and from which the coal had been removed. The process is then repeated and eventually encompasses the entire coal field, resulting in a series of ridges and valleys. More important, if not properly handled, the topsoil ends up on the bottom and the subsurface soil, often incapable of supporting plant life, is found on top. As a result, large areas could be rendered unsuitable for further plant growth and crop production. To minimize this potential damage, the federal Surface Mining and Reclamation Control Act requires that the topsoil be stored during the mining process and be used as the growing medium during reclamation. To date, over 1 million acres have been strip mined, causing several billions of dollars in damage to agricultural and other land.

Underground Mines. Underground (deep shaft) mines pose their own unique difficulties. The mining procedure is quite simple: Shafts are dug underground into a coal deposit, beneath the overburden. The coal-bearing materials are removed by railway car or conveyor belt (see Figure 6-8). As the mining progresses, pillars of coal are left standing to help support the overburden and minimize the potential for its collapse into the horizontal tunnel.

The environmental effects of deep mining are fourfold:

1. Much waste is brought to the surface which must be disposed. It is estimated that enough waste rock is generated in one year by coal

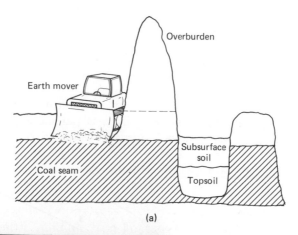

(a)

(b)

Figure 6-7 A typical strip mine:
(a) the strip mining procedure; (b)
strip-mined land (Sonora, Mexico).

mining to cover 1000 acres to a depth of 50 ft. This waste is generally piled in refuse banks. Some of this waste can be used as
structural fills and road bases, but such use is currently minimal.

2. The waste rock often contains pyrite (iron sulfide) impurities.
   When pyrite is subjected to weathering, it can react with water to
   form sulfuric acid in water. Acids can devastate aquatic life if they
   reach the nearby streams and lakes. This acid formation also occurs within the mine if proper precautions are not taken.

3. Refuse banks and abandoned mines often can ignite and create

Figure 6-8 A battery-powered tractor and trailer, articulated and highly maneuverable, carry coal from mining machine or loader to conveyor belt or rail loading point. (Courtesy of the National Coal Association.)

fires that are very difficult to extinguish. In addition to adding to our air pollution, coal fires can destroy a very valuable natural resource.

4. Abandoned mines can collapse or subside. In urban areas, streets and buildings can be destroyed. In rural areas, the paths of natural water streams can be altered and farmland ruined. Subsidence is a more prevalent problem when a fire has destroyed some or all of the coal pillars intended for overburden support. Backfilling of underground mines is possible and will reduce the subsidence, but it is quite expensive and is not usually done.

Another major difficulty created by underground mining is the inevitability of unhealthy mining conditions. Many miners die in cave-ins and explosions, in spite of the federal Coal Mine Health and Safety Act, which has, in recent years, dramatically reduced both the fatal and nonfatal accidents in coal mines. This law also requires very low levels of dust in the mines; thus it has reduced the incidence of black lung disease (a form of silicosis generated by the coal dust). In spite of our technological advances, however, coal mining is still a hazardous profession.

Combustion

When coal is burned, other environmental problems are created. Table 6-2 summarizes the percentages of the total amounts of the atmospheric air pollutants that arise due to coal combustion used to produce electricity (coal now is used to supply 45% of our electric power). Additional amounts of these pollutants are due to coal combustion which provides industrial process heat and industrial energy for other purposes. Also included in the table is a summary of the potential health consequences of these pollutants. These pollutants can also seriously harm materials and plants. Estimates of hundreds of millions of dollars in damage to agricultural crops are calculated annually.

When coal is burned, large amounts of soot are produced. This particulate matter can be a severe nuisance. As discussed in Chapter 3, the burning of most fuels, including coal, can enhance the greenhouse effect or, conversely, the extra particulates can shield the earth, lowering the average temperature. Simultaneously, the sulfur impurities that are present are converted to sulfur oxides. These sulfur oxides, especially when in the presence of particulate matter, can react to form sulfuric acid, one of the two major constituents of acid rain. Yet, removing large amounts of particulate matter intensifies the acid problem, because particulate matter is generally alkaline and therefore partially neutralizes the acid. Nitrogen oxides, also formed to a lesser extent in coal combustion, contribute to the acid rain problem. They react in the atmosphere to form nitric acid, the second major constituent of acid rain.

Acid rain is a serious threat to our environment. Acid rain attacks marble, limestone, concrete, and mortar, thus defacing buildings and monuments (see Figure 6-9). It has been estimated that acid rain has caused more damage to ancient Greek monuments since the 1920s than

TABLE 6-2   Coal-Fired Power Plant Pollutants

Substance	Total Pollutant (%)	Health Consequences
Sulfur oxides	75	Toxic, acid water, viral pneumonia
Nitrogen oxides	42	Irritants to breathing, discomfort to eyes
Particulates	21	Weather modification, restrict visibility, deposits in lungs or upper respiratory, dustfall and dirt
Hydrocarbons	2	Carcinogenic
Carbon monoxide	1	Heart disease, shortness of breath, retention of cholesterol

Source: Courtesy of the U.S. Department of Energy's American Museum of Science and Energy, operated by Science Applications, Inc.

Figure 6-9 Acid rain, a major problem resulting from coal combustion, corrodes marble, limestone, and mortar, leading to serious damage to buildings, monuments, and statues. In this picture taken at the Field Museum of Natural History, Chicago, the black crust consists of gypsum and soot. Behind crusts such as this, weathering continues, eventually causing the crusts to fall off, seriously damaging the structures. Since these crusts form only in protected areas, marble in sheltered regions can be more seriously damaged than that exposed to the rain.

was done in the preceding 2000 years. Acid rain also attacks plant life, and when it falls on lakes and streams, it can kill fish and other aquatic life. Over one-half of the lakes above 2000 ft in elevation in the Adirondack Mountains in New York, for example, are completely devoid of fish life, probably due to acid rain.

One of the major difficulties associated with acid rain is that it, and its effects, are not localized. It recognizes no regional, state, or national borders. Even the most remote areas of the world are not immune to its effects. The acid rain believed generated in the industrialized Midwest reaches the Adirondack Mountains and crosses our borders into eastern Canada (see Figure 6-10). The Scandinavian countries suffer seriously from the acid rain generated in the industrialized Western European countries. The political implications of this problem are obvious. Contributing to the difficulty is the uncertainty within the scientific community as to how acid rain is formed, how the pollutants are transported, which pollutants are actually involved in forming acid rain, and how much effect the watershed has on the formation of acid bodies of water. No consensus has been reached and no real solutions have yet been proposed.

Pollution Control Methods. It's an old and inadequate adage that "the solution to pollution is dilution." Sometimes electric power plants and industries are intentionally located in areas with sparse population

Figure 6-10 Acid rain, generated primarily by nickel-refining operations, devastated a large area around Sudbury, Ontario, Canada. Even decades later it has not been possible to reclaim this land.

and little other industry. For example, the Four Corners Power Plant, which serves Los Angeles, is located near the meeting point of New Mexico, Arizona, Colorado, and Utah, many miles away from the people it benefits. This policy of locating generating facilities great distances from those to whom they provide electricity raises many questions. Is it fair to have the environmental costs of power generation be borne by one group of people but the benefits realized by another? Is pollution really preferable in a low-population region? The health and materials damage is less, but the damage to the natural ecosystems may be greater. Yet it is also true that pollutants do occur naturally. It is the fact that they are emitted as a result of human activity in a concentrated form that causes difficulties. Thus many of the effects of pollutants can be minimized by either locating the plant in an area where little other pollution will be emitted and/or dispersing the pollutants by use of, for example, a "tall stack" (see Figure 6-11). However, as our society has become both more industrialized and more environmentally aware, this method of "control" is no longer considered adequate by many people.

Generally speaking, the best way to control a pollutant is never to generate it in the first place. This leads naturally to the preferred use of cleaner coal; coal with fewer impurities, particularly less sulfur. The sulfur content of coal generally ranges from 0.5 to 4%. In addition, coal contains large quantities of noncombustible materials, which must be collected and disposed.

Figure 6-11 Tall stacks are sometimes considered to be a solution to the pollution problem, since they disperse the pollutants over many miles. This is the 1250-ft stack at International Nickel Company of Canada, Ltd.

Coal low in sulfur and ash is not always available or economically feasible. Sometimes other factors also play a role. For many years Chicago factories, in spite of the area's dense industrial base, were required to burn high-sulfur Illinois coal.

Lower-grade coal can be cleaned prior to combustion to minimize the pollutants generated. The methods developed have primarily focused on sulfur removal. The efficiency of removal of sulfur varies with the type of coal and the form in which the sulfur is present.

The most commonly used method for control of sulfur and fly ash (the particulate matter, or ash, which is entrained in the exhaust gas stream) is stack gas cleanup: removal of the pollutants immediately

Figure 6-12 Baghouses such as these are very efficient in removing particulate matter. They often have an efficiency of 99.5%.

after combustion. Ash can be removed relatively easily by a variety of methods. Baghouses (Figure 6-12), which act rather like vacuum cleaners, and electrostatic precipitators, which utilize electric charge generated on the soot particles, are common methods and typically achieve greater than 99.5% efficiency.

The removal of sulfur is much more difficult. There are many possibilities but no single reliable, proven, and economic procedure. Most systems today involve the use of wet scrubbers, in which the sulfur is "washed" out of the airstream by a water/calcium carbonate mixture. Pulverized limestone or dolomite, both plentiful and cheap, can also be used to absorb the sulfur gases. Solutions of slaked lime or ammonia are other possibilities. All these methods have disadvantages and limitations.

PEAT

Peat, the lowest grade of coal, has been used as an inexpensive and easily available form of home heating fuel for many years in Finland, the USSR, and Ireland. Today a number of other countries, especially Sweden, are looking to their vast reserves of peat and considering large-scale utilization.

The form of peat traditionally used in households has been sod peat. This form (Figure 6-13) can be dug by hand or by machine. To-

Figure 6-13 Sod peat being harvested by hand (Connemara, Ireland).

day the industrial market is largely for finely milled peat. This finely granulated substance may be used as is, or compacted at pressures of 5 tons per square inch to form large rectangular briquettes with a typical heating value of 800 Btu/lb.

Peat bogs are usually, in their undrained state, 95% water. Even after initial draining, the bog remains 90% water; hence the peat must be dried before harvesting. To form sod peat, the material is turned three times with a plow and put into ridges. It thus dries to 35% water. Two sod peat harvests per year are possible. Milled peat is obtained by simply skimming a thin layer from the top of the soil, piling this material in long ridges in the middle of the field to dry. Up to 12 milled peat harvests can be obtained per year. "Sausage meat" peat is a relatively new invention by the Finns. The equipment, which can be drawn by an ordinary farm tractor, forms peat into rolls that look like large sausages. One continual worry is the potential loss of harvesting equipment. A large tractor, typically costing $50,000, could sink in minutes if it got bogged down.

Since the 1950s much of the peat from Finland, Ireland, and the USSR has been burned to produce electric power. In Ireland, for example, 20% of the electric power is obtained from peat. Coal boilers normally can be easily adapted to burn briquetted peat; however, milled peat requires special burners such as fluidized beds. The terminal building at Dublin Airport is just one of many buildings in Ireland heated by burning peat. In fact, peat is considered the most successful of Ireland's state industries.

Peat can also be carbonized to coke for metallurgical purposes, such as the production of steel. Peat from the north and west of Scotland is extremely reactive, and actually forms a better grade of coke than does coal, due to its lower levels of the impurities phosphorus and sulfur. Like coal, peat can also be converted to synthetic fuels. This approach is seriously under study in Sweden.

Although peat has been harvested mechanically since the 1940s, cheap coal and oil, and the promise of limitless nuclear power, reduced the likelihood of large-scale peat production until the early 1980s. This is perhaps unfortunate, for peat, in addition to being an indigenous fuel, has both economic and environmental advantages. In Scotland, for example, coal costs about $160 per ton; peat can be obtained as briquettes for about $35 per ton. In addition, after harvesting peat, the land underneath can easily be cultivated. Pilot studies have included growing grass on the land for hay production, or for drying, pelletizing, and subsequent use as winter feed for sheep. Growing biomass, especially wood such as willow, has also been attempted. The exact use of the cultivated land depends, of course, on the subsoil type. Typically, it does have a copper deficiency, which must be treated.

One of the major economic advantages of the peat industry, at least in the British Isles, is the impact it has had on the job market. Formerly, many Irish bog areas were very underdeveloped; there was little hope for the residents except emigration to England or America. Now this industry has created jobs. In Scotland's West Isles, where commercial peat harvesting is seriously under consideration, male unemployment levels were as high as 40% in 1982. It is estimated that for agricultural use one worker per 100 to 200 acres is required. In contrast, for peat production, the ratio is one person per every 25 acres. Government could not help but be interested in the promise of economic potential for these regions.

QUESTIONS

1. Discuss the question of "availability" of a fossil fuel.
2. Distinguish between underground shaft mines and strip mines. What are the typical characteristics of the resulting coal? Environmentally, what are the advantages and disadvantages of each method? How can these environmental effects be minimized?
3. Describe the historical change in patterns of the various fossil fuels. What historical events or discoveries could be associated with each of these changes?
4. What is acid rain? What are its effects? Which of its components can more easily be controlled? Why? Check your local environmental protection agency as

to methods designed to minimize acid rain formation. How do these regulations differ from current federal regulations?

5. Look into what plans the United States has for developing peat as a fossil fuel. Why haven't we done so already? What factors will hinder the growth of a peat industry?

CHAPTER 7

Petroleum: An Expensive Habit

Petroleum is vital to the U.S. economy. Automobiles rely almost totally on gasoline, derived from petroleum. In addition, most of the chemical feedstocks, fertilizers, and many pharmaceuticals are based on petroleum.

FORMATION

Although two theories exist as to the mechanism by which petroleum was formed, most geologists believe that petroleum and natural gas originated from living organisms rather than inorganic rock. The proposed types of organisms that may have served as sources range from low-order aquatic organisms (both fresh and marine) to land plants. Many questions still persist about the nature of the transformation process by which organic matter was converted into petroleum, how the petroleum migrates into reservoirs, and whether this process occurs in an early or late stage of petroleum production. It is known, however, that only 0.01% of the material produced each year by photosynthesis in the oceans would have to be converted to petroleum to account for the existing petroleum reserves in the world.

Current theory proposes that as the tiny marine organisms which lived millions of years ago in the shallow waters and near the shore died, their remains settled to the mud at the bottom of the oceans. Here

Figure 7-1 The drilling of oil wells such as this is a common sight in various parts of our country. (Courtesy of the U.S. Department of Energy's American Museum of Science and Energy, operated by Science Applications, Inc.)

bacteria caused the remains to begin to decay. With time, sediment drifted over the decaying plant and animal matter. As the sediments piled up, their tremendous weight pressed them into hard and compact beds of sedimentary rock. This weight simultaneously created heat and pressure, which, combined with bacterial action, converted the plant and animal remains in the mud beds into petroleum and natural gas. Tiny droplets of oil and bubbles of gas eventually migrated from the mud beds into the porous sedimentary rocks, which were usually sandstone or limestone. Less-porous sedimentary rock later covered the porous rock beds, sealing in the petroleum deposits. In later ages, many of the ancient seas were drained by movements of the earth's crust; dry land was formed above numerous of these deposits (see Figures 7-1 and 7-2).

PETROLEUM RESOURCES AND RESERVES

Like coal, the actual quantities of oil available in the future are uncertain. The answer to "How much oil?" depends on whom you ask and the methods used to make the estimate.

One common estimation method, the *finding rate method*, is to compare the quantity of oil discovered versus the number of feet drilled in exploration, then extrapolate to predict future probable discoveries. Figure 7-3 illustrates some typical results. It is easy to see that the "easy" oil generally was found a number of years ago.

Another, more complex method involves the *geological determination* of whether each step required in the petroleum formation process could have occurred at a particular location. These requirements include the likelihood of there being submerged shelves onto which

Figure 7-2 Oil pumps, employed once the well is drilled, do not have to be large, complex structures.

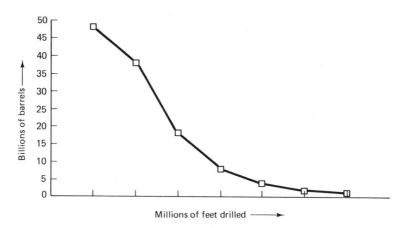

Millions of feet drilled ⟶

Figure 7-3 The finding rate method for estimating oil resources for oil in place.

marine animals could have been deposited, and whether it is possible that this area subsided. A second requirement is the nearby location of porous rocks to form reservoirs and the possibility of subsequent sealing of such areas by impermeable rock layers. Experimental evidence must be gathered by a variety of techniques, including seismic surveys, analyses of rock strata penetrated by prior drilling, rock outcrop maps, and records of drilling success.

Little agreement is found between these methods. For example, in west Texas, an area thoroughly drilled and explored and which has been producing oil for over 50 years, estimates as to the remaining oil vary by a factor of more than 2.

Which of the two methods described above is more accurate? The finding rate method is inherently conservative. It allows for no developments in geophysical surveying techniques, no unsuspected large fields, no major changes in the economics of petroleum. On the other hand, the subjective geological approach is more likely to be influenced by the overenthusiasm common to petroleum geologists. Which method is more accurate? With both methods, in some locations the estimates are high, in many others they have been too low. It is generally believed that actual drilling is the only way to be certain at the present time.

ENHANCED RECOVERY

Studies are currently being conducted concerning methods of *enhanced recovery*, methods used to extract more of the existing oil (and natural gas) from the earth's crust. For oil, the aim is to increase the recovery rate from 33% to 50% of existing petroleum. Five possible methods are being studied.

1. Flooding the oil-bearing rock with chemically thickened water, to "push" the oil out of the rock
2. Injecting chemicals called *surfactants* into the oil deposits to act as detergents and wash the oil out of the reservoir
3. Burning the heavier oil within the formation and using the heat generated by this combustion to drive the lighter oil, which is thinned further by the heat, to the production wells
4. Injecting steam or very hot water into the reservoir under pressure, using the heat to thin the oil in order to increase the flow rate
5. Injecting carbon dioxide into the oil-bearing rock to dissolve in the oil, making the oil lighter so that it would flow more readily to the wells

Similar procedures are also being tested for natural gas, except here the primary approach is to break apart the submerged rock formations that retain the gas. The use of chemical explosives and the pumping of a water and sand mixture under high pressure into the rock formation are two methods under study.

The purpose of these enhanced recovery techniques is not to reverse the downward trend of U.S. gas and oil production or to make the United States energy self-sufficient, for those goals are impossible by this method. They can, however, extend U.S. production to allow time for the development of other fuel resources. But the use of these technologies is expensive and they can require significant amounts of energy. If steam flooding is used, the equivalent of about one-third of the oil extracted is needed to drive the steam generators alone. In-situ combustion burns about 10% of the oil present and generates corrosive gases. Furthermore, compressed air is required to maintain this controlled combustion, and the compression costs may be a major consideration. Similarly, the chemicals used to lower the surface tension of the oil and polymers to thicken the water can be very costly, and often their use is not justified economically. Oil and energy prices are therefore a critical factor in decisions regarding the use of these recovery procedures.

LOCATION OF THE DEPOSITS

The locations of the major world petroleum deposits are summarized in Figures 7-4 and 7-5. The Middle East is the major source of petroleum. There are only several hundred wells located in this area, but each well can produce tremendous amounts of oil. Although the United States

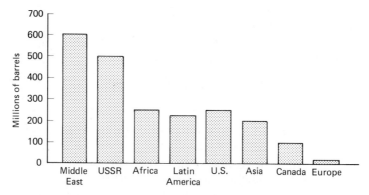

Figure 7-4 Location of major petroleum deposits. (Courtesy of the U.S. Department of Energy's American Museum of Science and Energy, operated by Science Applications, Inc.)

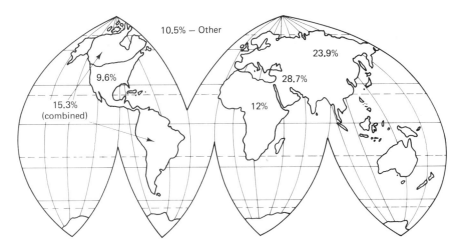

10.5% — Other

23.9%

9.6%

28.7%

15.3%
(combined)

12%

Figure 7-5   World distribution of oil.

has over 600,000 wells, the average U.S. well produces less than 20 barrels of oil per day, whereas one well in Saudi Arabia produces over 10,000 barrels per day. Most wells in the Middle East average thousands of barrels per day. Since the major cost of oil production is the drilling of the well, it is clear how the oil from the Middle East can be produced much less expensively—for a few cents per barrel—than can domestic oil. Oil from Alaska, for example, costs over $12 per barrel to recover.

As with coal, Western Europe has very little petroleum. For many years it has had to import virtually all the oil it needs. The recent discovery of North Sea oil, between England and Norway, may, however, change the picture somewhat for countries that border that body of water.

As observed in the case of coal deposits, the USSR has a plentiful supply of petroleum. Although it is an exporter of oil, many of these deposits are located in the more remote regions of the country, so extraction may be economically prohibitive. The United States possesses very small oil reserves compared to the Middle East. Yet we still have significantly more oil reserves than some of the other industrialized nations. Our relatively large supplies of oil have allowed our society to become very dependent on "cheap" petroleum to power our automobiles, as well as many of our industries.

In the United States large deposits of petroleum are located in Pennsylvania, Texas, Oklahoma, and Alaska. Seventeen percent of the oil supply in the United States is located offshore (see Figure 7-6), near the coasts of Louisiana, Texas, Florida, and California.

Figure 7-6 Offshore drilling provides 17% of the U.S. oil supply. (Courtesy of the U.S. Department of Energy's American Museum of Science and Energy, operated by Science Applications, Inc.)

USE PATTERNS

The primary uses of petroleum in the United States are summarized in Figure 7-7. Figure 7-8 lists the typical form in which petroleum products are sold.

The first oil in the United States was discovered in Titusville, Pennsylvania, in 1859 at a depth of only 69.5 ft. There was little demand for the new fuel except as a substitute for whale oil, which was then used in lamps. This changed dramatically with the advent of the gasoline-powered automobile in the early twentieth century. U.S. petroleum use peaked during the mid-1970s when we obtained approximately 40% of our total energy from oil. Since then, percentagewise, our consumption of petroleum has declined slightly.

How long will our petroleum last? Clearly, the answer to that question depends on our rate of petroleum consumption. The rate of

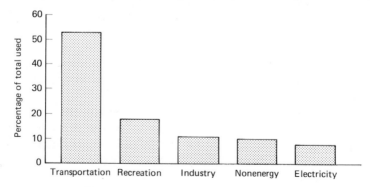

Figure 7-7 Primary uses of petroleum.

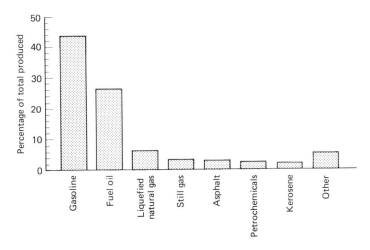

Figure 7-8 Typical distribution of petroleum products. [From James L.
Pyle, *Chemistry and the Technological Backlash* (Englewood Cliffs, N.J.:
Prentice-Hall, Inc.), © 1974, p. 100. Reprinted by permission of Prentice-
Hall, Inc. Data from *Kirk-Othmer Encyclopedia of Chemical Technol-
ogy*, 2nd ed. (New York: John Wiley & Sons, Inc., 1972), Vol. 15, p. 78.]

growth in consumption is best described by an exponential model—that
is, the rate at which the consumption increases is dependent on the
actual consumption at that time. Under these conditions, the doubling-
time calculation discussed in Chapter 4 is appropriate. Even if the increase in consumption is only about 2% per year, roughly equivalent to
the U.S. growth in population, every 35 years the rate at which we consume oil will double. If we maintained our current rate of consumption,
it was estimated by the Department of Energy in 1980 that the supply
could meet U.S. needs for only another 21 years. The world supply is
significantly greater, but if U.S. consumption increased as it has in preceding years, even that oil would be depleted in less than a century,
assuming that the United States had access to all the world's supply of
oil! Needless to say, to think that would be a possibility is sheer fan-
tasy. In fact, in the future the United States is likely to be able to
obtain a smaller portion of the total world oil supply. Not only will the
importing of oil become economically unfeasible, but the developing
countries will, themselves, require a much larger share of the total to
get and keep their industries producing.

Today, one-sixth of U.S. electricity is generated by burning oil. This
is expensive, unnecessary, and generally considered to be an inappropriate use of a very valuable fuel. Figure 7-9 illustrates the geographical
regions with a heavy dependence on oil for electricity generation.

Approximately one-half of the petroleum we use is now imported.

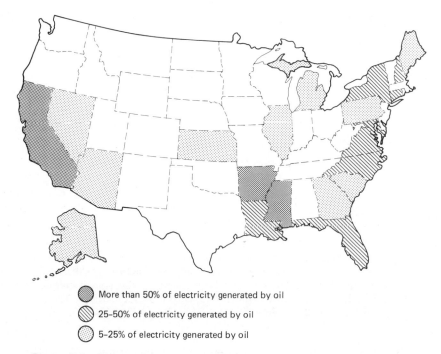

More than 50% of electricity generated by oil

25–50% of electricity generated by oil

5–25% of electricity generated by oil

Figure 7-9 States with a heavy dependence on oil for electricity generation. (Courtesy of the Society for the Advancement of Fission Energy.)

In 1979, the United States used 17.0 million barrels daily, 8.2 million of which were imported primarily from the Middle East. Crude oil imports in 1979 were almost 80% above the 1973 level prior to the OPEC oil embargo.

In the past, U.S. dependence on imported petroleum was not the case. In 1960, the United States produced 33% of the world's supply whereas the Middle East produced only 25%. However, by 1972 the United States produced only 18% of the world's supply and the Middle East produced 35%. One of the major reasons for this rapid shift was the larger Middle Eastern supply, resulting in much lower production costs.

What is likely for the future in terms of petroleum use? The experts are predicting significantly lower use in the United States. The Department of Energy changed the 1977 estimate of 83 million barrels per day (mbd) for 1990 to 52 mbd in its 1979 estimate. Even more interestingly, Middle East oil production estimates (for 1990) decreased from 61 mbd to 26 mbd in the same time period. Clearly, the declines in world consumption have been predicted to be primarily at the expense of the large Middle Eastern exporters.

OPEC

In 1960, a number of countries, primarily in the Middle East, banded together to form the Organization of Petroleum Exporting Countries (OPEC). The member states are Saudi Arabia, Iraq, Kuwait, United Arab Emirates, Qatar, Libya, Algeria, Venezuela, Indonesia, Nigeria, Gabon, and Equador. The purpose was twofold: (1) gradually to obtain ownership of the foreign investments in their countries, and (2) to control price. Until 1973, the petroleum market was really a "buyers' market." Oil prices were extremely low (Middle Eastern oil could be purchased for $1 per barrel with transportation costs contributing an additional $2 per barrel). The OPEC embargo of 1973 changed that situation. The OPEC countries totally eliminated exports to both the United States and Netherlands (Shell is a Dutch company) and reduced exports to other countries. The immediate result was a 1000% price increase, from $1 to $10 per barrel. Domestic oil simultaneously went from $3 to $9 per barrel, with a corresponding 60% increase in profits. Since that time, oil prices have increased by another factor of 4 to 5.

Our continuing dependence on petroleum, particularly imported petroleum, has meant that we must pay these high prices, which have been a major contributor to worldwide inflation.

DEPLETION ALLOWANCES

Within the United States there are also a number of political considerations regarding petroleum production and use. One of these considerations involves petroleum depletion allowances. For many years depletion allowances have provided economic incentives for the further exploration and use of many natural resources, including petroleum. In calculating income subject to taxes (either personal or corporate) the owners of any mineral (or petroleum) production operation are allowed to deduct from their gross income an "allowance," supposedly to compensate for the fact that the capital asset, the minerals, are being depleted. The depletion allowance is similar in concept to capital depreciation, although the details differ as to how it is computed.

These depletion allowances have encouraged exploration for new petroleum supplies and increased use of petroleum. They provide no incentive for conservation of this very valuable fuel and, as such, are contradictory to much of the current U.S. energy policy.

ENVIRONMENTAL CONCERNS

The difficulties associated with extraction of oil arise primarily with offshore wells. Blowouts, such as the one that occurred in the Gulf of Mexico in the summer of 1979, can foul beaches, kill marine birds, destroy breeding grounds, and concentrate pesticides such as DDT since such substances are 100 million times more soluble in petroleum than in water.

Similar concerns over oil spills occur during the transportation of petroleum, often by large supertankers. Oil spills on water from either source are particularly difficult to handle, for oil can form a very thin layer 0.01 inch thick over a 25-square-mile area in 8 hours. The fish breeding grounds, particularly in the far northern areas, are considered too valuable a resource to endanger. For countries such as Norway, potential danger has resulted in a decision to limit oil exploration in the North Sea, particularly above the 62°N parallel of latitude. Destruction of these spawning grounds could seriously disrupt the traditional life-styles and livelihoods of a large percentage of their population.

The long-term effects of oil spills are unclear. Recent studies of the many oil-laden German and U.S. ships sunk during World War II have led to the hope that the impacts may be rapidly minimized by natural forces. It may be that large spills are not as detrimental as wide-spread, constant small discharges, such as those emitted by oil tankers.

Actually, the number of offshore oil well blowouts is quite small. Of the 20,000 offshore wells drilled by the United States since 1947, only two have resulted in a spill that reached shore. Until the very large and well-publicized blowout of the Mexican offshore drilling rig at Ciudad del Carmen in June 1979 (the oil of which did reach the eco-logically sensitive Texas coast), there had not been any large oil spills due to offshore drilling since 1972. During that same time period, how-ever, there had been many sinkings of supertankers off the coast of the United States and other countries which resulted in large oil spills.

There are a number of methods which are being developed to con-trol the oil spill problem. Surface-film-forming chemicals can minimize the oil spreading or even drive the oil back into a thicker layer. Liquid organic gelling agents are being studied with the hope that one could be used to gel the crude oil in the cargo hold of a distressed tanker. If the mixture would escape from the ship, it would float in a big mass. Poly-urethane foam, to absorb spilled oil, appears feasible. The foam can be regenerated by mechanical squeezing, and the oil can be 98% recovered.

The creation of a vortex, such as that generated by stirring a glass of water, will draw an oil slick into the center from where the oil can be removed by pumping.

Alaskan oil production also has created many concerns. Oil was discovered in 1968 on the North Slope, near Prudhoe Bay. An 800-mile-long pipeline was built from there to a tanker base in an all-weather bay, Valdez, located in the southern part of Alaska. This pipeline (Figure 7-10), a marvelous engineering feat, built through a land rich in wildlife and scenery, provides 1 to 2 million barrels of oil a day. The potential dangers of the pipeline include (1) pollution due to leaks and breaks; (2) the movement of the relatively warm oil in the pipeline, and the additional frictional heat generated by the flow of the heavy crude, could melt the permafrost, which might lead to collapse of the pipeline supports, resulting in more leaks and breaks; (3) the very slow return of all vegetation due to the short growing season; and (4) the disruption of wildlife migratory paths, although "bridges" have been included to minimize this problem.

Combustion of petroleum creates many of the same difficulties as those of coal combustion. The processing of the crude petroleum, however, does remove a good share of the sulfur, so the products (gasoline, heating oil, diesel fuel) burn more cleanly. This minimizes the potential hazard of acid rain.

Figure 7-10 Construction of the Alaskan oil pipeline. (Courtesy of the U.S. Department of Energy's American Museum of Science and Energy, operated by Science Applications, Inc.)

PETROLEUM PROCESSING

Drilling

Petroleum processing starts at the point of discovery. There is no one method used to locate the precise spot to drill. Modern exploration methods include magnetic, gravity, and seismic (shock) measurements.

A *gravimeter*, or gravity meter, measures the attraction due to the earth's gravitational field. Since different types of rocks affect the magnitude of the gravitational field in different ways, if many gravimeter readings are taken over a wide area, it is possible to obtain maps of the shape and depth of the various rock layers.

A *seismograph* measures shock waves. An explosive charge is set off within a shallow hole. The shock waves that are generated travel downward until they reach an underground layer of hard rock, from which they are reflected. The length of time it takes the waves to reach that underground rock layer and return is an indicator of the depth of the layer.

A *magnetometer* records the differences in the magnetic effects of various rocks. There are always local variations in the earth's magnetic field. These variations are caused by the type of rock structure below the surface. Magnetometers can be trailed behind an airplane, permitting the traversing of large areas and the development of large-scale geologic maps. Even with these techniques, however, there is no precise way to guarantee whether oil is present short of actual drilling.

Land Drilling. Oil wells can be drilled by two methods, the (rather outmoded) cable-tool method and the rotary method.

Cable-tool drilling equipment consists of a heavy bit attached to the end of the cable. The bit is successively lowered and raised, crushing the soil and rock. Occasionally, water is flushed into the hole. The resulting mixture is removed periodically, by bailing.

Rotary drilling is by far the more efficient method to drill a well. The drill pipe consists of a steel pipe attached to a bit with sharp teeth. The exact design of the bit depends on the nature of the formation through which the well is being drilled. As the well gets deeper, additional sections of pipe are added to that already in the hole, permitting greater depths to be reached. Mud, from a mud pit located beside the well, is pumped through the drill pipe and bit into the well for cooling and for coating the walls to prevent cave-ins and to keep out subsurface water. Lignosulfonates, from waste spent liquor generated in the pulping of wood for papermaking, are often mixed with the mud to act as a binder. This is a clear example of the waste for one industry serving as a valuable raw material for another. If the underground formation is very soft or if it contains much water, the well must be lined with steel pipe.

This casing is lowered into place and the inside is then filled with wet cement. A plug is placed on top of the cement, then mud is forced down on the top of the plug. This additional pressure forces the cement down and then up the outside of the casing, where it is allowed to harden. The plug and mud are removed, and drilling can continue using slightly narrower drills.

Once the well has reached the oil-bearing formation, it is necessary to be very cautious so that a wasteful "gusher" does not form. The mud column usually exerts enough pressure to hold back the oil flow until a system of control valves is placed on top of the casing. The mud can then be removed. The oil and gas reaching the surface are separated in a tank, and any water is removed from the oil, which is then ready for refining.

Offshore Drilling. The offshore drilling procedure itself is very similar to that of land drilling. The major difference is in the support systems for the drilling apparatus.

In land drilling, a huge steel framework, or derrick, is set up over the spot where the well is to be drilled. The derrick is typically 100 to 200 ft tall, and supports primarily a system of pulleys and blocks which are used to hoist the drilling equipment into and out of the well.

Three types of systems are concerned with the discovery and recovery of petroleum from under the ocean floor.

1. Drilling ships are engaged mainly in the discovery of oil. Indistinguishable from any ocean-going cargo vessels except for their drilling derricks, their purpose is to drill boreholes in areas where seismological and geological studies suggest that oil may be present.

   Although these vessels are equipped with the most technologically advanced navigational equipment, it is very difficult for them to maintain their position in heavy winds and waves, and under those circumstances drilling must be suspended.

2. For the actual drilling and recovery of oil, the "jack-up" system is currently in almost universal use. It consists of a platform equipped with a drilling rig, cranes, helipads, and living accommodations. The platform is firmly anchored to the ocean floor by steel legs. It, like the drilling ship, has drawbacks. It is not very mobile. It can be used only in relatively shallow water (typically in no more than 450 ft of water) due to the cost of extending the length of the legs. It must suspend operations in even moderately heavy weather. Moreover, the safety of such rigs is a question. Two have already been lost in the North Sea, with the consequent loss of many lives. The suspected cause is either design, structural, or metallic faults in one or more of the legs.

3. The newest type of oil rig is the "tension leg drilling/production" or "semisubmersible" platform. The huge platform, in the shape of an equilateral triangle, can be constructed of a special ferro-concrete rather than of steel. The steel legs and a variety of horizontal and diagonal cylinders are hollow, providing tremendous buoyancy overall. Extended from each leg are massive steel cables which can be attached to anchors on the ocean floor. This setup permits the rig to operate at depths greater than 1000 ft, hence far out on the continental shelf.

There are other distinct advantages to this type of drilling rig. Its buoyancy puts the anchor cables under tension sufficient to eliminate the pitching and rolling of the platform due to heavy seas. Drilling need not be suspended during severe storms. The system is also mobile, for to move it requires only lifting its anchors and relocating.

Refining

The first step, after receipt of the raw petroleum, is refining. Crude oil contains a number of inorganic impurities which are detrimental to the operation of the refinery units. Chlorides can react with water to form corrosive hydrochloric acid, which would lead to equipment deterioration. Sand and other suspended matter would cause plugs. Salt would cause scale buildup in heat exchangers. Other substances may prevent the proper chemical reactions from occurring. These impurities must first be removed.

Petroleum is a complex mixture of various products which can initially be separated on the basis of their relative volatilities, that is, by how easily they are evaporated. As the temperature of the mixture is gradually increased, the various components are successively vaporized and collected in a fractionating column. Dissolved gases such as methane, the propanes, and the butanes are the first to vaporize. The methane is sold as natural gas; the latter two are liquefied under pressure and are sold as bottled gas.

The next substance vaporized is petroleum ether, a mixture of quite volatile liquid components. Gasoline distills when the temperature reaches 50 to 200°C. Gasoline itself is a complex mixture, but the size of the molecules is larger than those previously vaporized.

After the gasoline, kerosene, then fuel oil and diesel oil are vaporized. The only slightly volatile waxes and lubricants are last to vaporize at reduced pressure. Left are the nonvolatile asphalt-type residues and tarry materials which are produced during heating.

Cracking

From year to year and even from season to season, there are differing demands for these products. Until the automobile was popular, kerosene was in greater demand than gasoline; now, of course, the reverse is true. In the summer, gasoline consumption rises; in the winter, heating fuels are more desirable. Regardless of which of these substances is the prime demand, further refining is always required to convert less desirable to more desirable products.

Presently, much effort is expended to convert larger molecular substances, such as kerosene, to gasoline. Crude petroleum contains no more than 18% gasoline. This conversion of one petroleum product to another is usually done by a procedure known as *cracking*. Cracking is accomplished by heating the products in the presence of solid catalysts such as silicas. Not only does this create smaller molecules, suitable for use as gasoline, but it also produces bulk chemicals which can be converted to specific pharmaceuticals, pesticides, or synthetic fibers.

Although cracking can be used on all larger molecules, typically those containing 12 to 20 carbon atoms per molecule are split into smaller (6 to 12-carbon-atom) molecules with a volatility suitable for gasoline. In addition to producing smaller molecules, cracking also produces large quantities of ethylene and other *unsaturated hydrocarbons*. Unsaturated hydrocarbons are those substances whose molecular structures contain one or more carbon–carbon double bonds or triple bonds. Because of these bonds, these types of molecules tend to be fairly reactive. These products are useful in enhancing the antiknock properties of gasoline, and they can be used as intermediates for the production of other chemicals. Ethylene is probably the most valuable of these petrochemicals, for it forms the basis for the production of polyethylene for packing materials, ethylene oxide which is used in epoxy resins, ethylene glycol for antifreeze, and butadiene, used widely in the manufacture of synthetic rubber.

The original technique of cracking consisted of exposing a flowing stream of, for example, fuel oil to high heat and moderate pressure in steel tubes. The resulting material was then released to a large chamber at lower pressure, where the gases and liquids could separate. The condensed vapors and liquids could then be further separated by normal fractional distillation. Although this thermal cracking method is still used in some smaller refineries, in most cases it has been replaced by catalytic cracking.

Catalytic cracking involves both high temperatures and the use of a catalyst. Natural catalysts are composed primarily of silica and alumina, with small amounts of other materials. Synthetic catalysts are similar to the zeolites used in home water softening. The purpose of the catalyst is to provide a solid surface on which the desired chemical reaction can occur. Frequently, petrochemical reactions will occur more rapidly under these conditions. For this reason the catalysts are designed to have a tremendously large surface-to-volume ratio. The catalyst particles are both very small and very porous. Typically, their surface areas are measured in terms of acres!

After catalytic cracking to split apart the larger molecules, *catalytic reforming* may be done. The purpose of this step is to form aromatic compounds from straight-chain and cyclic hydrocarbons. Aromatic hydrocarbons are cyclic but possess alternative single and double carbon bonds. An example would be the conversion of methyl cyclohexane to toluene:

Toluene is very beneficial in gasoline, for it increases the octane number without the use of lead.

In addition to the pollution potentially generated in the drilling and transportation of the petroleum, refining and cracking also require water for processing. Subsequently, it becomes polluted and must be treated. The treatment methods are generally multistage procedures designed to recover and regenerate the wastes as much as possible to form products low in sulfur. The quantities of wastewater can be quite large: The Standard Oil Complex at Texas City, Texas, for example, produces 23 million gallons of polluted water daily, equivalent to that generated by a city of about 300,000 people. The water must be treated by some biological treatment method so that microbes can digest and remove the petroleum wastes before it is released to the environment.

QUESTIONS

1. Why is enhanced recovery an important possibility? How is it accomplished? What are the major current limitations as to its use?

2. Compare the finding rate method for estimating oil resources and the geological approach. Which do you prefer? Why?

3. Explain why some petroleum is much more costly to recover than that from other deposits.

4. Why is the importing of oil a political issue? Consider this question from a variety of perspectives.

5. Summarize the environmental effects of oil use. What methods are used to minimize each of these effects? Consider all stages in its production and use.

6. Describe the procedure for cracking petroleum. What are the goals? Why is it necessary?

7. What methods are used for recovery of oil in offshore deposits?

CHAPTER 8

Natural Gas: The Ideal Fuel?

Natural gas deposits are found together with petroleum deposits, and areas rich in one are generally rich in the other. In past years, natural gas was considered a nuisance since it was almost impossible to store, difficult to transport, and can be fatal if too much is inhaled.

NATURAL GAS RESOURCES AND RESERVES

How much natural gas is available? Until the early 1980s it was believed there was much less natural gas than any of the other fossil fuels. If we were to rely on U.S. supplies alone, experts believed that our reserves would not last to the beginning of the twenty-first century. It was estimated that these reserves would last 10 to 15 years if we continued to consume them at the present rate, and about 4 to 5 years if we kept increasing our demand to the rate experienced during 1970s.

Since then, additional evidence indicates that our natural gas supplies may be significantly larger than previously supposed, possibly even large enough to last until solar and other new technologies can make a significant impact on our energy supplies. Two separate reasons are responsible for this new position concerning natural gas supplies: one scientific, the second political.

The scientific reason is based on the processes believed involved in the original formation of the fossil fuels. As discussed in Chapter 7, the current theory as to fossil fuel formation is based on the fact that at one time or another, every part of the earth's surface has been covered by ocean water. The plankton and other one-celled organisms fell to the ocean floor when they died. Over millions of years these layers of fossilized sediment became thousands of feet thick; the resulting heat and pressure generated by their sheer weight led to the conversion of the organisms to the hydrocarbons that make up our fossil fuels.

The specific depth within these layers at which the plant and animal matter was located is an important factor in determining specifically what fuel was formed. To depths of about 15,000 ft, both oil and gas can be formed. At greater depths the heat generated is so intense that the more complex oil compounds are not stable, and they decompose to form natural gas. The result is two distinct layers, one with both oil and natural gas, and the second, deeper layer, with only natural gas.

During the earlier days, geologists drilled only for oil and the natural gas was wasted. Moreover, drilling technology was essentially limited to depths of less than 15,000 ft, primarily because of a lack of ability to identify probable gas locations at any greater depth. Over 90% of the drilling operations were futile. The invention of computerized seismometers during the 1970s now allows geologists to "see" to depths of 40,000 ft, better pinpointing the areas to drill. As a result, new gas fields are being located in the Rockies (an area near the border of Wyoming and Utah is now believed to have more gas than Alaska's North Slope), near Baton Rouge, Louisiana, and in an area near the Texas and Oklahoma border, to name only a few. Furthermore, it is believed by some petroleum geologists that these may only be the "tip of the iceberg."

What does this mean in terms of supply? It would depend somewhat on which estimates one assumes to be correct for the supplies remaining. A low estimate (2.4×10^{15} Btu or 2.5×10^{18} J) would mean a 120-year supply at the present rate of consumption. Even if we increase our consumption somewhat, this would indicate that we can possibly use natural gas, which is considered clean and relatively environmentally benign, until there is further development of other energy technologies.

The second reason for the change in predictions regarding natural gas supplies is a combination of three "political" factors. The 1976–1977 heating season was the coldest in 45 years; thus natural gas supplies dwindled and even ran out in some locations. Simultaneously, the oil and gas industry was attempting to get Congress to deregulate gas prices on the basis that the proven reserves (those actually drilled into and measured) had dropped drastically, from a 14-year supply in 1968 to a 10-year supply in 1977. The industry lobbyists insisted that low, controlled prices discouraged exploration. At this time, the current U.S.

president was also trying to sell the public on the need to conserve energy—that there was an "energy crisis." He was so adamant in his belief that our oil and natural gas supplies were "running out" that two high-level government officials who disagreed with his interpretation of the supply data, the head of the U.S. Geological Survey and an official in the Energy Research and Development Administration, were both fired.

In October 1978, Congress passed a law slowly phasing out regulation of natural gas prices. Almost immediately an excess of natural gas reached the market. Although the Secretary of Energy explained it as a temporary "bubble" in the pipelines, this bubble persisted, due presumably to increased financial incentives for the exploration and to the new seismometer developments.

LOCATION OF THE DEPOSITS

The locations of the major natural gas fields are summarized in Figure 8-1. Much of the natural gas consumed in the United States has been supplied from Canada and Mexico. From these neighboring countries, it is possible to transport natural gas in the gaseous state, by pipeline to our supply network. Additional natural gas has been imported from the Middle East, particularly from Algeria (see Figure 8-2). When natural gas is transported by supertanker, it must first be cooled to −259°F to liquefy it, producing *liquefied natural gas* (LNG), which reduces its volume to a small fraction of what it would occupy as a gas. Handling LNG is dangerous, however, for it possesses quite an explosive potential

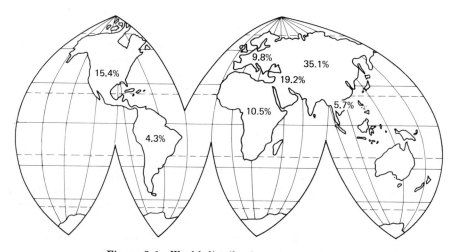

Figure 8-1   World distribution of natural gas.

Figure 8-2 Large supertankers are used to transport natural gas to the United States, particularly from Algeria. (Courtesy of the U.S. Department of Energy's American Museum of Science and Energy, operated by Science Applications, Inc.)

because it boils easily and burns readily. Many seaports have been unwilling to have LNG supertankers dock and be unloaded.

USE PATTERNS

Natural gas has been known for many centuries. A shepherd in ancient Greece who noticed that his sheep were acting strangely at a certain place on the mountainside investigated and discovered a seepage of vapors which were thought to be the breath of the god Apollo. A temple was soon built at that place, Delphi, which rapidly became the religious center of all Greece.

Several early civilizations discovered that natural gas could be used as a fuel. The Chinese were the first to use it industrially. Nearly 3000 years ago, they obtained gas from 2000-ft-deep wells, sent it through bamboo pipes, and used it to evaporate brine to make salt.

Natural gas was discovered in the United States as early as 1775 in West Virginia. Pioneer children enjoyed setting fire to seepages in locations such as along Lake Erie. The first commercial well was not dug, however, until 1821 in Fredonia, New York. Although gas manufactured from coal had been used for lighting purposes since the early nineteenth century, and the city of Baltimore lighted its streets with gas in 1816, it was not until 1824 that natural gas was used commercially for lighting purposes in the town of Fredonia. By 1850 use of natural gas for street lighting was widespread in the United States. It was not until after 1879, when Thomas Edison invented the electric light, that its use for lighting began to diminsh. But by then its practicality as a cooking fuel had been demonstrated.

Today over four-fifths of all gas used in the United States is natural gas. Natural gas, which was once thought a nuisance associated with oil drilling and was often flared to dispose of it, is now one of our most valuable fuels.

Currently, there are adequate supplies for small-scale uses such as home heating, although natural gas use continues to be phased out for large-scale industrial users whenever possible, largely due to recent increases in price.

THE POLITICS OF NATURAL GAS

The prices for natural gas had been regulated in the United States for many years. The federal government had the authority to control interstate prices, so they maintained it at artificially low levels, while intrastate prices were higher. Gas suppliers, understandably, preferred to sell within the states where the natural gas was produced. This led to large supplies and excesses in states such as Texas and Oklahoma, while in many other areas there were sporadic and sometimes acute shortages.

This was changed in 1979 with the National Energy Act. This Act equated the inter- and intrastate prices; raised prices, especially for the largest consumers; and mandated the total rescinding of all price controls on natural gas by 1985.

Because natural gas travels in pipelines, natural gas supplies, rather like electricity, can be considered a monopoly. Heating oil can be brought by any supplier; natural gas can be furnished only by the company that owns and maintains the piping. This has encouraged a proliferation of governmental regulations regarding supplies, prices, and use of natural gas.

There have also been other politically caused problems regarding natural gas use. Prior to 1973, it was assumed that natural gas use could be greatly increased in the future. The Clean Air Act was based on this assumption. The Clean Air Act of 1970 authorized the establishment of nationwide primary and secondary air standards, thus limiting the emissions from both old and new industries. The particular substances of concern include SO_2, CO, and NO_x, hydrocarbons, particulates, photochemical oxidants, and many other hazardous substances. Because natural gas burns so cleanly, especially compared to coal, industry supposedly could greatly decrease its emissions by converting to natural gas use. However, soon after many industries invested millions of dollars converting their processes to natural gas fuel, the oil embargo of 1973 forced many governmental policymakers to realize that natural gas supplies, in particular, were limited to the short term. Suddenly, coal was back in favor. The industries that had recently installed natural-gas-handling equipment had to switch back to coal-burning facilities. Simultaneously, they had to install very costly collection equipment so that they could meet pollution control guidelines.

EXTRACTION AND PROCESSING

Natural gas wells are drilled in much the same way as oil wells; in fact, oil and natural gas are often found in conjunction with one another. As the well is drilled, a casing is driven down to protect the hole. The casing keeps out water and sand and provides a path along which the natural gas can rise.

When a deposit is tapped, the gas rushes out under high pressure. As the gas is extracted, the pressure drops until it is so low that it cannot be removed in sufficient quantities for economical operation. To prolong the life of the well, a percentage of the gas can be recycled back into the well. This gas gathers additional gas from the sand and porous rock. Water may also be pumped into a gas field to maintain the pressure.

Enhanced recovery methods are now under considerable study. Since the natural gas deposits are encapsulated very tightly in rock formations, these enhanced recovery methods are aimed at splitting apart the rock formations. Two methods show promise:

1. The use of chemical explosives
2. Pumping a sand and water slurry into the crevices in rock strata in the gas field

Once a well begins to produce, it is capped to prevent the loss of natural gas. A pipe transports the gas first to a dehydration unit, where the moisture is removed, then to an extraction unit, where impurities are removed. The impurities may be valuable by-products, for they include hydrogen, methane, ethane, propane, butane, heavier hydrocarbons, and hydrogen sulfide. The methane and hydrogen are not readily absorbed by an oil; hence the other materials may be removed by passing the mixture through a low-temperature, high-pressure adsorption column.

The methane–hydrogen mixture then enters a compressor, where it is raised to a high pressure for transport. As the gas travels through the pipelines, it will be recompressed approximately every 100 miles. Water-cooled condensers may also be used to lower the temperature of the gas, for the compression simultaneously raises both the temperature and pressure.

The network of natural gas pipelines in the United States is the nation's greatest privately owned land transportation system. There are almost twice as many miles of gas pipeline as of railroads, and two to three times as many miles as of oil pipelines.

Since the nation's demand for natural gas will vary with the season of the year, and since a pipeline cannot accommodate radical changes in demand, during the summer many gas companies store large amounts of gas near the larger cities for use during the cold winter months. The underground storage areas may be formations that once produced oil or gas, but are now dry, or they can simply be rock formations covered by hard rock which will seal in the gas once it is pumped in. In this way, hundreds of billions of cubic feet of natural gas are stored annually.

ENVIRONMENTAL CONSIDERATIONS

Compared to the other fossil fuels, the environmental effects of natural gas are relatively benign. The sulfur levels are very low and the ash that results from combustion is essentially nonexistent. Natural gas "spills" mix readily with the atmosphere and diffuse. The only real considerations are:

1. The combustion does produce carbon monoxide, carbon dioxide, and nitrous oxides.
2. Like oil wells, natural gas wells could subside and they can be considered aesthetically displeasing.
3. The development of a large natural gas deposit in a particular area could cause secondary environmental effects associated with the growth of the resulting population center. This would potentially have a major impact only if the area under development had been sparsely populated, such as much of the West and Alaska.

QUESTIONS

1. Discuss the factors that probably led to the upward revision of our estimated natural gas supplies.
2. Discuss the statement, "Natural gas supplies can be considered a monopoly." What regulations are there in your community to govern natural gas suppliers?

3. Distinguish between the types of rock strata in which petroleum and natural gas are located.

4. A major natural gas deposit was just discovered in an uninhabited part of the U.S. Southwest. What type of environmental effects would there be due to the growth of a nearby town which houses primarily the workers and their families?

CHAPTER 9

Derived Fuels:
An Expensive Alternative

There are a number of fuels that can be derived from the fossil fuels. These new fuels have several advantages. In some cases synthetic fuels (synfuels) are more readily adaptable to a variety of combustion techniques (coal liquefaction and gasification) than the raw materials from which they are derived. In other cases they can be used to extend our reserves of a scarcer fuel, such as petroleum (oil shale, tar sands). Frequently, in the conversion process, sulfur and other impurities are removed, creating a much cleaner burning fuel and minimizing the detrimental effects.

COAL GASIFICATION AND LIQUEFACTION

Our large reserves of coal can be converted to gaseous and liquid forms, to relieve our dependence on petroleum and natural gas. Solid coal, even when pulverized, cannot, for example, be generally used to power an automobile. Coal typically has a heating value of 12,000 to 14,000 Btu/lb; gasoline averages about 20,000 Btu/lb. An automobile engine would have to have major modifications if it were to burn not only a solid substance, but also one with substantially less energy per pound than gasoline. (A new technological advancement now permits the substitution of simple coal–water mixtures for heavy fuel oil in boilers, however.) In the process of converting the coal, many of the impurities, such as sulfur and ash, are removed simultaneously. This produces a much cleaner burning fuel than conventional coal. However, much of the potentially available energy is lost. The current processes are only about 50 to 75% as efficient as burning the coal directly.

There are a number of technologically feasible ways of producing these "synthetic fuels" and they are not new. In South Africa, for example, there has been commercial production of a liquid fuel from coal since 1954 by Sasol (see Figure 9-1). Germany had 19 synthetic fuel plants operating in 1942. The reason the United States has not moved in this direction is simply one of economics. Until the escalation of petroleum and natural gas prices during the 1970s, there was no economically justifiable reason to produce such fuels. Large government investments such as the $20 billion dedicated for the period 1980–1984 were directed at discovering more efficient and economical methods. The proposed plans called for the development of a synthetic fuels industry 20 times larger than the German effort during World War II.

Figure 9-1 Sasol synthetic fuels plant, South Africa. (Courtesy of Sasol/Fluor.)

The Synfuels Process

Figure 9-2 illustrates with a flowchart a typical procedure used to produce liquid and gaseous fuels from coal. A comparison of the molecular structures of solid, liquid, and gaseous fuels illustrates the basic principle. Coal is almost pure carbon, containing only small amounts of hydrogen (and, of course, other impurities); in petroleum, the carbon-to-hydrogen ratio is approximately $1:2$; in natural gas, the ratio is $1:4$. As the amount of hydrogen increases, the molecules comprising the fuel become smaller and the fuel exists as a liquid or gas rather than a solid at normal temperatures. Moreover, the energy available during combustion increases. (Coal has a heating value of about 13,500 Btu/lb; gasoline, 20,000 Btu/lb; methane, 23,860 Btu/lb.) Hence in the liquefaction

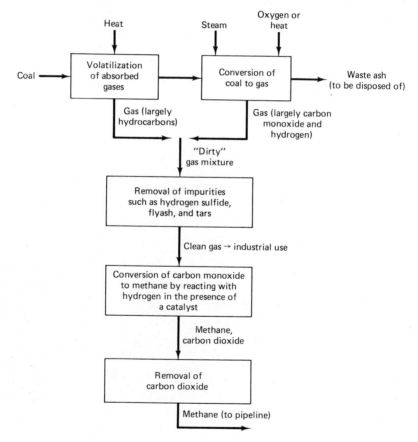

Figure 9-2   Typical procedure for coal gasification.

and gasification process it is necessary to increase the hydrogen level. An ideal overall reaction is simply to add coal to water, producing a reaction that forms methane and carbon dioxide. This, however, cannot be accomplished in one step, and therefore a sequence of several reactions in separate steps are required.

Other possible conversion schemes exist. Pyrolysis, heating coal in the presence of its volatile products, produces both liquid and gaseous fuels. Liquid fuels can also be produced by dissolving the coal in an appropriate solvent, adding hydrogen, then properly adjusting the physical parameters, such as temperature and pressure, to produce the desired product. Many modifications and schemes are now being tested experimentally, aimed primarily at improving the economics of the process. Pilot facilities, such as the synthane coal gasification plant near Pittsburgh (Figure 9-3), themselves can be major investments. This plant, built in the 1970s, cost $15 million and was designed to process 72 tons of coal per day into 1.2 million cubic feet of pipeline-quality gas.

Figure 9-3 Pilot coal gasification plant. (Courtesy of the U.S. Department of Energy's American Museum of Science and Energy, operated by Science Applications, Inc.)

ECONOMICS OF COAL GASIFICATION/LIQUEFACTION

Currently, there are many methods available for the production of the various synthetic fuels. Most of these methods are, however, chemically nearly identical to one another. The economics of the various production methods are therefore also quite similar, although methane is inevitably more economical to produce than fuel oil.

Regardless of the particular process chosen, the efficiency is typically about 67%: the products contain only about two-thirds of the energy originally present in the coal raw material. Under this assumption, with current technology, the cost of producing these fuels is typically $5.60 to $6.30 per million Btu for methane and $40 to $44 per barrel ($6.60 to $7.40 per million Btu) for fuel oil (1983 dollars). During this period, when OPEC crude oil was selling for $29 per barrel, the purchasing price, as delivered to the customer, of conventionally produced natural gas and fuel oil were, respectively, approximately $5.80 per million Btu and $40 per barrel. Although these costs appear to be approximately the same, it is important to note that the prices for conventional fuels are the net cost to the customer, whereas the synthetic fuel costs are only the production costs, a portion of the total costs.

The effect of process efficiency is not a major factor in determining these costs. Increasing the efficiency from what might be considered to be the minimum acceptable efficiency, 50%, to the maximum efficiency thermodynamically possible, 75%, would save only about $5.90 per barrel for fuel oil and a comparable amount for methane. If this increase in efficiency would also require higher investment costs, as is likely, there might be no economic benefit at all.

The major developmental work, therefore, is now focusing on methods to reduce the capital investment costs without decreasing the efficiency. The newer plants are expected to require 20% less capital investment than the best existing commercial facilities. This would lead to a decrease in overall production costs of about 10%. Hopes for much lower prices in the future are unrealistic unless major scientific breakthroughs are made. In-situ coal gasification processes, for example, have shown some encouraging preliminary results but have yet to demonstrate any significant economic advantages.

OIL SHALE

Much of the shale rock located near the junction of Utah, Wyoming, and Colorado is impregnated with heavy oil. This oil, once extracted, can be refined by methods similar to those used to obtain oil from conventional petroleum reserves.

The potential quantity of petroleum available from oil shale is tremendous. This one deposit alone, the Green River deposit, is estimated to hold 1.8 trillion barrels of oil. The U.S. Department of the Interior has estimated that 600 billion barrels could be recovered from this deposit over the next several decades with existing technology. The estimated recoverable oil in the deposit is nearly equal to the total present proved conventional crude petroleum worldwide, and more than 20 times the proved U.S. conventional reserves.

There are two major environmental difficulties, however, associated with shale oil production.

1. Much of the oil shale is located in an area where there is very little water. Three gallons of water are required to produce 1 gallon of oil from shale resources. Thus, even a relatively small, 1 million barrel per day plant needs one-half of the potential water supply in the area. Furthermore, the deposits are in areas of low population, negligible industry, and good air quality. Shale oil development would undoubtedly lead to increased commercial and industrial development, which would put further pressures on the water supply.

2. Shale oil production also produces tremendous quantities of waste rock (Figure 9-4) because 85% of the mined material is waste rock.

Figure 9-4 Waste rock produced by shale oil extraction. (Courtesy of the U.S. Department of Energy's American Museum of Science and Energy, operated by Science Applications, Inc.)

It is estimated that to obtain the same amount of energy from oil shale as from coal, it would be necessary to mine five times as much shale as coal, resulting in 17 times as much waste material.

There are two methods that can be used to produce oil from shale. The shale can be strip mined (by power shovel, front-end loader, etc.) and the rock then treated with heat and pressure (in the absence of air) in a large retort to release the heavy crude oil. The waste rock can then be used to backfill the areas that have been mined, but this is often difficult. Even if this is done, the generation of air pockets by the process of retorting means that a larger volume of rock would remain than was removed. The rock that cannot be backfilled could leach and release materials that might contaminate nearby streams and lakes.

In-situ extraction is another production possibility for oil shales. Here no external retort is required. A shaft is drilled into the deposit. At least 25% of the shale in the area around the shaft is removed. Explosives are used to generate heat and pressure underground, releasing the crude oil, which flows downward and is collected in conventional wells. This process does have advantages in reducing the quantities of waste rock produced, but is not technologically perfected yet, and questions exist as to its seismic effects, due to the expansion of the rock strata.

Many oil companies have conducted large-scale tests to determine the economic feasibility of extracting shale oil. For example, Rio Blanco Oil Shale Company, a general partnership of Gulf Oil Corporation and Standard Oil Company (Indiana), conducted both in-situ and surface retort tests. It was hoped that evaluation of the two technologies would be completed and commercial production begun by the late 1980s. However, this particular project has been "mothballed" for the present. The company expects oil shale development to be resumed at a future date, but using a modification of the existing technologies.

TAR SANDS

Tar sands such as the deposits found in large quantities along the Athabasca River in Alberta, Canada, can provide a very heavy crude oil similar in quality to that available from oil shale. Commercial mining and extraction of this Canadian deposit began in 1967. The sand must be mined, washed in hot water to remove the oil, and the sand must then be returned to the mine. These tar sands are approximately 15% petroleum (by weight); thus almost 7 tons of sand must be mined to extract 1 ton of oil. This process creates a solid waste problem similar to that found in oil shale extraction. The main difference, however, is that they are located in a region that has abundant water supplies. Small deposits of tar sands also exist in the United States, but, to date, there is no reliable survey as to their exact location or the estimated quantities of oil they contain.

OTHER DERIVED FUELS

Synthetic fuels can also be produced from a variety of other organic-based substances. Wood and wood waste products, including waste paper mill liquors, can serve as a raw material to produce a variety of fuels. For example, these wastes are a potential source of alcohol. Municipal solid waste can also serve as *refuse-derived fuel* (RDF). Studies are currently under way studying particular plant species that contain oil. For example, euphorbias, which grow in desert areas, can produce 10 to 20 barrels of oil per acre, from about 10% of the plant, in a 6 to 7 month growing season. Sorghum and sycamore are other possibilities, but their growth would compete with food production. These plant-derived fuels are discussed in Chapter 14.

QUESTIONS

1. What advantages are there in converting a solid fuel such as coal to a liquid or gaseous form? What disadvantages are there?
2. What chemical basis underlies all coal-to-synthetic fuel conversion schemes?
3. Contrast the two methods for shale oil recovery outlined in the text. Which do you favor? Why?
4. If a major deposit of tar sands were discovered near the U.S. oil shale deposits, would there be any major advantage(s) to mining the tar sands instead of the oil shale? Why or why not?

CHAPTER **10**

Electric Power: The Light of Our Lives

Most of our common appliances are powered by electricity. Electric energy also lights our buildings and helps run our industries. Who uses it? What is this useful form of energy? How is it generated? What can it really do for us?

The electric power companies generally classify their customers as:

Residential
Commercial
Industrial
Street and highway lighting
Farm and ranch

The relative amount that each demands is illustrated in Figure 10-1. In this figure, farm and ranch use is included with industrial.

In recent years, the demand for electrical power has grown very rapidly. A major reason is simply its convenience. For the consumer, it can produce high-quality lighting; it is relatively safe, clean, and quiet; and it is versatile and dependable.

To understand what electrical power really is, together with its benefits and limitations, it is necessary to consider some of the elementary concepts associated with electrical energy.

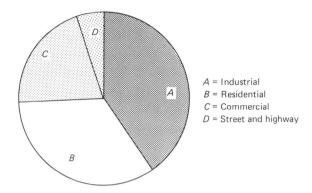

Figure 10-1   Classification of electric energy users. (Courtesy of the American Association for Vocational Instructional Materials.)

BASIC ELECTRICAL TERMS AND RELATIONSHIPS

Electric current consists of a flow of electrons. Electrons are subatomic particles, each of which possesses an electrostatic charge of 1.6×10^{-19} coulombs (C). There are three basic terms that are used to describe electrical energy. *Voltage* (V) is the electrical potential that causes electrons to flow through a material. The units of voltage are volts (V). One volt is the equivalent of 1 J of energy per coulomb of charge flowing. The *current* (I) is the number of electrons flowing through a material in a given period of time. It is measured in amperes (A). One ampere corresponds to a flow of 1 C of charge per second past a fixed point in the circuit. Each coulomb of charge is comprised of 6.25×10^{18} electrons. *Impedance* (Z) is the opposition to the flow of electrons through a material. It is measured in ohms (Ω). The impedance of a circuit depends on the size, length, temperature, and the composition of the material through which the electrons travel. These parameters are related by the equation $V = IZ$.

Electrons flow readily through metals such as copper and aluminum (called *conductors*), but only very slightly through materials such as ceramics, plastic, and paper (called *insulators*). When electrons flow, they follow a wire or path called a *circuit*. Circuits are generally made of conductors which are protected by insulation. An *open circuit* (Figure 10-2a) is one that is (intentionally or accidentally) broken, preventing electron flow. An open circuit can be created by opening a switch. *Short circuits* (Figure 10-2b) allow the current to flow through only part of the circuit. Usually these are accidental, and they may be unsafe.

There are two forms of electric current: direct current (dc) (Figure 10-3a) and alternating current (ac) (Figure 10-3b). A *direct current*

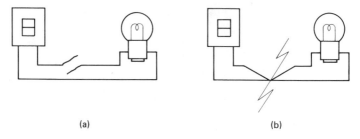

Figure 10-2   Types of circuits: (a) open circuit; (b) short circuit.

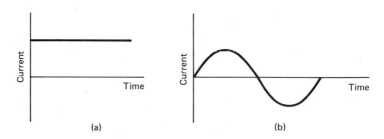

Figure 10-3   Types of current: (a) direct current; (b) alternating current.

occurs when the electrons flow in only one direction in a circuit, whereas with an *alternating current* the electrons flow first in one direction and then in the opposite direction. Direct current is what is supplied by a battery such as that used in a flashlight, a hand-held calculator, or to start an automobile. Once a circuit is established, one terminal of the battery will always supply electrons and the other will always accept electrons, at a steady rate. The magnitude of an alternating current, on the other hand, will vary with time. The current will start from zero, build to some maximum value, decrease again to zero, then increase to the same maximum, but with the electron flow in the opposite direction. In contrast to direct-current circuits there is never a net flow of electrons through a wire with alternating-current circuits. The electrons will move for a short period of time in one direction, but an instant later they will reverse their direction and flow an equal distance in the opposite direction. Since alternating-current systems have some technological and economic advantages, it is the form usually supplied by electric power companies.

A distinguishing characteristic of alternating current is its frequency. When the current first flows in one direction, reverses, and then is ready to flow again in the original direction, one *cycle* is completed. The number of cycles per second is called the *frequency*, measured in hertz (Hz) (that is, 1 hertz = 1 cycle per second). In the United States and most other countries 60 Hz is standard, although a few countries use 50-Hz current.

A power line from an electric power company is a circuit consisting of two components: a phase wire and a ground (or neutral) wire. A *ground wire* is a conductor that is connected electrically to ground or earth. Ground, or earth, is theoretically at 0 V or what is called zero potential. Therefore, the ground/neutral wire is also at 0 V. The *phase wire*, on the other hand, is a conductor that has some energy difference, voltage, with respect to ground. Most circuits in a home, for example, have a ground wire at 0 V and a phase wire with a voltage of 120 V. (Household stoves are usually an exception, in that they have wires at +120 V, 0 V, and −120 V.)

Electrical Energy and Power

Electrical energy (E), measured in joules, is the product of voltage × current × time (VIt). Since $V = IZ$, energy can also be calculated by I^2Zt or V^2t/Z. Rather than discuss energy itself, the term *electrical power* is often used. Power (P) is the energy per second available from an electric current. Power is measured in the units joule per second, which is known as a watt (W). Power can thus be considered to be the rate of use of electrical energy and can be calculated by dividing the energy by the time ($P = E/t$).

Power is also calculated using the relationship power = voltage × current ($P = IV$). The watt is therefore equal to the volt-ampere (see Chapter 2).

Appliances, light bulbs, and so on, are rated according to their power consumption in watts or kilowatts (kW), where 1000 W = 1 kW. Electricity usage, however, is measured using watt-hour meters, which really are energy meters. The power company charges for the *energy* used, or the rate at which energy is consumed (the power) multiplied by the time it was consumed ($E = Pt$), not the power itself. For example, a 100-W bulb consumes 100 watt-hours of energy in 1 hour. This is equivalent to 0.100 kilowatt-hour (kWh), or 360,000 J of energy.

Consider the two cases in which power is supplied to a 4800-W air conditioner. If the air conditioner is served at 120 V, the normal household voltage, $I = 4800$ W/120 V = 40 A. On the other hand, if it is served at 240 V, $I = 4800/240 = 20$ A.

What is the difference, besides the obvious difference in current? Which would be preferable? Any wire will overheat and set fire to its insulation and surrounding material if too much current flows through the wire. The maximum safe current for a wire of fixed cross-sectional area is called the *current-carrying capacity* of the wire. The capacity of a conductor is rated in amperes. In other words, a wire can carry only

TABLE 10-1   Conductor Capacity

Size (AWC)	Diameter (inches)	Capacity (amps)	Cost (per foot)
14	0.06408	15	$0.088
12	0.0808	20	$0.128
10	0.1019	30	$0.212
8	0.1285	40	$0.390

Source: Wisconsin Public Service Corporation.

a certain number of electrons each second without overheating and being damaged. The current capacity depends on the diameter of the wire. Higher amperage requirements thus lead to larger wire diameter requirements, and larger-diameter wires cost more. Table 10-1 lists some typical wire capacities and costs. To serve the air conditioner at 120 V requires a No. 8 wire at $0.039 per foot; at 240 V requires only a No. 12 wire at $0.128 per foot.

The load of a circuit is the total of the appliances, lighting, and other service which must be powered by that circuit. It is equivalent to the sum of individual power requirements and is thus measured in watts. As noted previously, power can be calculated by voltage times current. Voltage can similarly be defined as the product of current times impedance. Therefore, load (or power) can be defined as the product of the current squared times the impedance ($P = I^2Z$).

$$P = VI$$

$$V = IZ$$

$$P = (IZ)I = I^2Z$$

For example, consider the circuit in Figure 10-4. A 200-W light bulb, served from a 120-V circuit, requires 1.67 A ($I = 200$ W/120 V). Or we could reverse the calculation and calculate the impedance of the bulb:

$$V = IZ$$

$$Z = \frac{V}{I} = 120 \text{ V}/1.67 \text{ A} = 71.9 \ \Omega$$

$$P = I^2Z = (1.67)^2 \times 71.9 = 200 \text{ W}$$

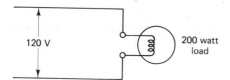

Figure 10-4 Circuit with a 200-W light bulb.

Transformers

A transformer is a device that is used to change voltage (for example, 14,400 V to 240 V). Simultaneously, it correspondingly changes the current. A transformer consists of two coils, adjacent to or overlapping but insulated from one another electrically (Figure 10-5). The voltage supplied to one coil is transferred to the other by magnetic coupling, generally by using an iron core, as illustrated in Figure 10-5, but modified by the number of turns of wire in the two coils: $N_1/N_2 = V_1/V_2$. If, for example, we wish to convert 14,400 V to 240 V and the former is applied to a coil with 1000 turns of wire, the second coil must consist of

$$\frac{14{,}400 \text{ V}}{240 \text{ V}} = \frac{1000 \text{ turns}}{x \text{ turns}}$$

$$x = 17 \text{ turns}$$

The power must remain constant or a transformer would violate the principle of conservation of energy. Consequently, the current must vary inversely with the voltage: $V_1 I_1 = V_2 I_2$. The efficiency of a transformer is quite high, about 98%. Transformers are exceedingly important in the transmission and distribution of ac current and are used to both increase and decrease the voltage values and current values.

Voltage Drop.   As the current passes through the various loads in the circuit, the voltage "drops," or decreases, because the electrons are

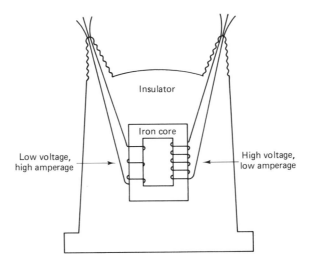

Figure 10-5 Principles of a transformer.

losing energy. The amount of the voltage drop is equal to the product of the current times the impedance of that load, $V = IZ$. To maintain a desired voltage (for example, 120 V) it is necessary at various points in the circuit to use a transformer which will slightly boost the voltage. This is not a consideration in most small circuits but is of considerable importance to a power company, where governmental regulations state that everyone must be serviced with a minimum of 114 V.

Power Losses.   A voltage drop produces a power loss. This loss is consumed by the current flowing in both the phase and neutral wires. Normally, the resistance of the wire itself is ignored compared to the other quantities of electricity that are involved in the circuit. Long power lines, however, require minimizing these power losses for economic reasons.

The power loss on an electric line can be calculated using the relationship $P = I^2Z$. To minimize the power loss, it is essential to minimize the current, I. This is accomplished by using high voltage and low current whenever that power needs to be transmitted over long distances. For this reason, electricity typically is generated at 2200 V and then raised by a transformer to 345,000 V, 138,000 V, or 115,000 V for transmission to major metropolitan load centers (or other utilities), then lowered again for distribution to homes and industries. The ability to avoid large power losses in long-distance transmission is the major reason alternating current is more economical. Direct current cannot be easily altered.

ELECTRICITY PRODUCTION AND DISTRIBUTION

Generation Methods

How is electricity originally generated? Electric power generation is all similar, even though different methods are used to provide energy to rotate the turbine. Generating electricity consists of converting another form of energy to electrical energy. Usually, the original energy is in the form of fossil fuels or nuclear energy. It can also be some form of mechanical energy, such as moving water or wind. Figure 10-6 summarizes the sources of energy used for electrical production in the United States.

Figure 10-7 illustrates the basic design of either a fossil fuel or a nuclear generating plant. The fuel provides heat to convert liquid water to high-pressure steam. This steam contains fast-moving molecules that if properly directed can cause a turbine to rotate. Water, as steam, occupies a much larger volume than does a similar amount of liquid.

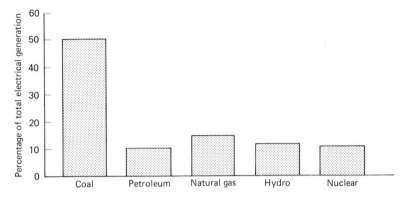

Figure 10-6 Source of fuel for electrical power generation in the United States. (Courtesy of the U.S. Department of Energy's American Museum of Science and Energy, operated by Science Applications, Inc.)

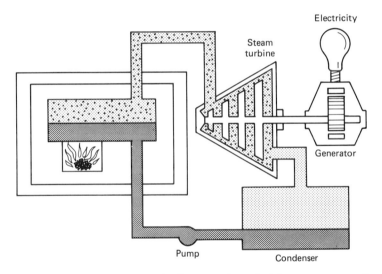

Figure 10-7 Typical schematic diagram of an electric power generating plant. (Courtesy of the U.S. Department of Energy's American Museum of Science and Energy, operated by Science Applications, Inc.)

In expanding, the steam exerts energy on its "container." This energy can be tapped by the turbine blades (Figure 10-8), causing the turbine to rotate. Initially, the fast-moving molecules in a hot gas are moving in random directions and will not cause rotation of turbine blades. The turbine blades respond instead to a flow of material. The random motion of the molecules must therefore be converted as much as possible to linear flow by a specially designed nozzle before the desired rotation will occur. The turbine is connected to an electrical generator (Figure 10-9),

Figure 10-8 Turbine blades that convert steam energy to electrical energy.

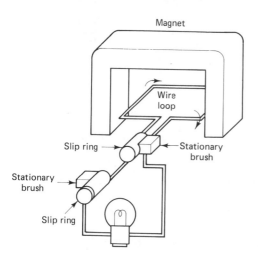

Figure 10-9 Simple alternating-current generator.

which consists of a wire-wound core placed between magnets. The rotating shaft on the turbine causes this core to rotate within the magnetic field. Any wire moving in a magnetic field experiences a force that causes the electrons in the wire to move, producing an electric current. The movement of the electrons is then transmitted to the place of use (i.e., homes, businesses, and industries). Hydroelectric plants are similar, except that moving water provides the mechanical energy to propel the turbine. The energy source is really immaterial to the electrical power generation; any convenient and reliable source is suitable.

Efficiency. Electrical power generation is not a very energy efficient process. In fact, some of the energy potentially available is lost at each step in the process. Figure 10-10 lists typical efficiencies at the various steps.

Costs. How much does this all cost? The cost will vary from one locality to another, and may fluctuate significantly, but typical values for a system using coal, nuclear, or hydro are capital costs ranging from $600 to $1000 per kilowatt of capacity and fuel cost of $85 per kilowatt of electricity generated.

Transmission and Distribution

After the electrical energy is generated, it must be transmitted and then distributed.

Transmission. The transmission system transmits electrical energy from generators to local load centers through wires supported by structures and insulators (see Figures 10-11 and 10-12). Table 10-2 summarizes the types of transmission and their typical construction costs. In transmitting power, two cost factors must always be considered. They are the initial costs, which are greater when higher voltages are used, and the power losses, which are greater when the lower voltages are used.

Substations. A substation (Figure 10-13) is a group of transformers. The major purpose of a substation is to reduce the circuit voltage from the transmission level to the distribution level. As discussed previously, not only is the voltage altered by a transformer, but so is the current; the power is what remains constant. If, for example, a transmission voltage is dropped from 115 kV to 25 kV for distribution, and the load (power) is 10,000 kW, the transmission current of 87 A (10,000 kW/115 kV) is increased simultaneously to 400 A (10,000 kW/25 kV). Costs of substations range from $350,000 to $500,000 depending on size.

Distribution. The distribution system distributes electric energy from the substation to customer service transformers located on poles through wires and transformers supported by poles and insulators.

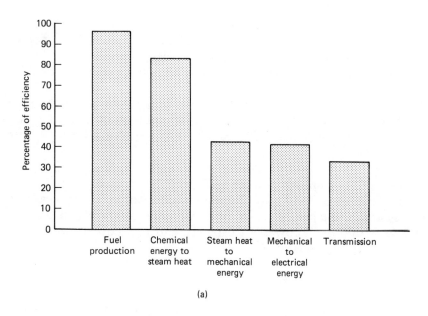

(a)

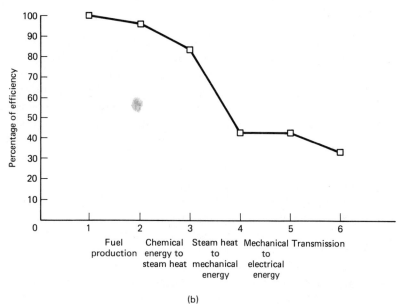

(b)

Figure 10-10 Efficiency of electricity production. (a) Efficiency of each step; (b) overall efficiency. (Courtesy of the U.S. Department of Energy's American Museum of Science and Energy, operated by Science Applications, Inc.)

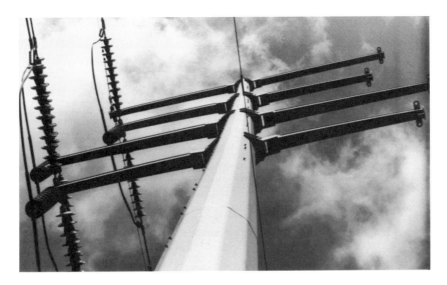

Figure 10-11   Transmission wires and poles such as these are a common sight all over the United States. (Courtesy of the Wisconsin Public Service Corporation.)

Figure 10-12 High-voltage transmission lines require special handling. (Courtesy of the Wisconsin Public Service Corporation.)

Typical primary voltages are 4160 V, 12,470 V, and 24,940 V. The secondary system distributes the energy from a customer service transformer to a customer's meter. The likely voltage will depend on the nature of the customer. Residential users typically require 120/240 V, commercial users 120/208 V, and industrial users 277/480 V or 2400/ 4160 V. (See Figure 10-14).

TABLE 10-2   Typical Types and Costs of Transmission
Construction (in 1984 dollars)

Types	Costs
Interconnections to other utilities	
345 kV	$160,000/mile
138 kV, 115 kV	90,000/mile
Service to major metropolitan load centers	
138 kV, 115 kV	90,000/mile
Service to smaller load centers	
69 kV, 46 kV	60,000/mile

Source: Wisconsin Public Service Corporation.

Figure 10-13   Electrical power substation. (Courtesy of the Wisconsin
Public Service Corporation.)

Regulators.   The voltage requirements are established typically
by state codes. These minimum (and maximum) values must be adhered
to by the power company. Because of voltage drops and power losses
along the line, it is necessary at various points in the distribution
system to install voltage regulators to maintain these minimum values.
A regulator is a variable transformer that will automatically maintain
a desired output voltage. A regulator can change the input by ±10%.

Figure 10-14 Pole-mounted transformers lower the voltage from transmission to distribution levels. (Courtesy of the Wisconsin Public Service Corporation.)

It typically has a control which can be programmed to maintain a constant output voltage based on the amount of load being served.

Examination of a light-load and a peak (maximum)-load situation can illustrate this. Figure 10-15 illustrates the variation in demand as a function of time for residential, farm, and industrial users, respectively, and the overall demand. Industrial loads are the only ones that are relatively constant, typically distinguished only by holidays and weekends. Assume a peak load of 5000 kW and a light load of 1000 kW. If the load is serviced at a nominal 14.4 kV, and the wire impedance is 4 ohms (Ω), the voltage drops are 1388 V and 278 V, respectively.

$$P = IV_{\text{nominal}} \qquad V_{\text{drop}} = IZ$$

$$5000 \text{ kW} = I \times 14.4 \text{ kV} \qquad V_{\text{drop}} = 347 \text{A} \times 4 \ \Omega$$

$$I = 347 \text{ A} \qquad\qquad = 1390 \text{ V}$$

$$1000 \text{ kW} = I \times 14.4 \text{ kV} \qquad V_{\text{drop}} = 69.4 \text{ A} \times 4 \ \Omega$$

$$I = 69.4 \text{ A} \qquad\qquad = 278 \text{ V}$$

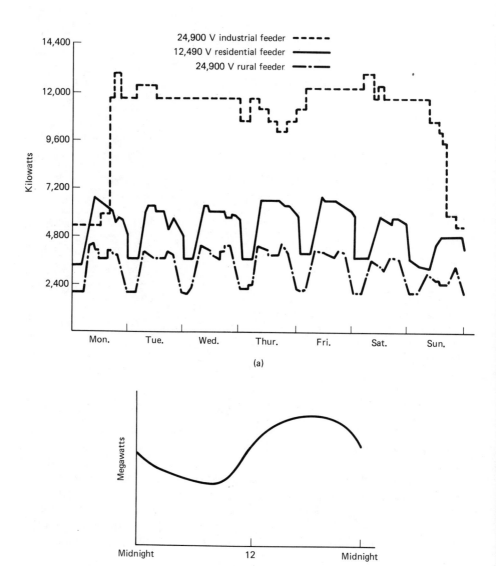

(a)

(b)

Figure 10-15 Energy demand: (a) industrial, residential, and rural. (Courtesy of the Wisconsin Public Service Corporation.) (b) Overall energy demand. (Courtesy of the American Association for Vocational Instructional Materials.)

The voltages at the load are thus actually 13,000 V (90% of nominal) and 14,100 V (98% of nominal).

$$V_{nominal} = 14.4 \, kV = 14,400 \, V$$

$$V = V_{nominal} - V_{drop}$$

$$= 14,400 \, V - 1390 \, V = 13,000 \, V$$

$$= 14,400 \, V - 278 \, V = 14,100 \, V$$

The peak-load case, 90% of nominal, is probably beyond specifications. A regulator can, however, raise the peak voltages by 5%, thus raising this value to an acceptable voltage.

Meeting Varying Electricity Demands

Electricity generally cannot be stored; it must be used as it is generated. But electricity demands vary seasonally (Figure 10-16), as well as with the time of day, as illustrated in Figure 10-15. The variations in demand cannot be totally met by the use of voltage regulators; thus the power generated must also vary. This is accomplished by using base-load and peaking-power generation facilities. *Base-load* plants generate the majority of the electricity. Their power output is generally constant and they provide adequate electricity to meet demand at minimum consumption levels. *Peaking-power generation facilities* are easily started and stopped. Hydroelectric plants, internal combustion turbines, and heavy-duty jet engines are frequently used for this purpose. Peak demand can also be satisfied by the purchase of electric power from other utilities. All power plants are interconnected by power grids, allowing power suppliers to assist each other when the demand is greater in a given area. Typically, electricity can be economically transferred several

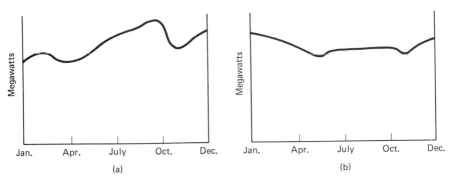

Figure 10-16 Seasonal variation in energy demand. (a) Warm climate; (b) cold climate. (Courtesy of the American Association for Vocational Instructional Materials.)

hundred miles. This purchased energy is more expensive, however; thus its use is avoided if at all possible.

Peak-load generating facilities are also generally more expensive to operate than base-load facilities. Thus if the load can be stabilized, there are definite economic advantages for both the power company and its consumers. For this reason, large-scale users, such as industry, are frequently charged lower rates for off-peak electricity use (for example, at night). In many cases, the rates charged large users are based on the total energy consumed during a given period and on the maximum demand generated, even if that demand occurred for only a very brief time (for example, 30 seconds in a 1-hour period).

Peak-load generating facilities are also generally more expensive to operate than base-load facilities. Thus if the load can be stabilized, there are definite economic advantages for both the power company and its consumers. For this reason, large-scale users, such as industry, are frequently charged lower rates for off-peak electricity use (for example, at night).

In many cases, the rates charged large users are based on the total energy consumed during a given period and on the maximum demand generated, even if that demand occurred for only a very brief time (for example, 30 seconds in a 1-hour period).

Many industries generate their own electrical power, as well as process steam. Controlling the fraction of each produced can aid in minimizing any sudden and short-term increases in purchased electrical power demand. "Peak-shaving" can also be accomplished by *load shedding:* temporarily reducing or totally eliminating some of the load. Load shedding can be of many forms. It may consist of reducing the air conditioning and water chilling in summer or lowering the heating in winter. It may mean reducing a grinding load in a pulp mill or postponing a melting operation in a foundry. The criteria used to select what systems to be shed usually include the capacity to be removed or added, the present and projected needs for the resources, and the lengths of time each has been in or out of service.

Individual consumers may also be charged "time-of-day" rates, depending on their geographical location. In this way, it is possible to level the demand curve somewhat.

HYDROPOWER

The energy of falling water can be harnassed to produce electric power. Hydropower is one of the oldest power sources. At one time a large number of small dams were used for electrical production. Currently, the Pacific Northwest and the Tennessee Valley depend heavily on hydroelectricity (see Figure 10-17). At the present time, only 4% of the

Figure 10-17 Norris Dam in Tennessee, the first of the TVA dams. (Courtesy of the Department of Energy's American Museum of Science and Energy, operated by Science Applications, Inc.)

U.S. budget is supplied by hydropower. It is estimated that even if we greatly increased our efforts in hydro generation, the total contribution will not be significantly greater than it currently is because of the lack of appropriate sites. These sites are located where water is swiftly moving through a narrow gorge (see Figure 10-18). The river is generally dammed, creating a reservoir and flooding the land behind the dam. Obviously, land areas near larger cities are not a good choice to flood. Most of the unused "suitable" sites are rich in scenery and wildlife and may be parts of National Parks. Hydro dams can also block the spawning efforts of fish, such as salmon. This obstacle can be only partially alleviated by the use of fish ladders as shown in Figure 10-19, so the populations of desirable fish are often reduced. When considering hydro capability, it is necessary to evaluate the need for more electricity in light of the associated environmental consequences. The choice will often depend on people's values.

But even assuming that we chose to develop all, including marginally appropriate, hydro sites, hydropower could not become a major source of energy in the United States. There is simply not enough hydro potential energy available. If the gravitational potential energy of every drop of rain that falls on the United States each year were converted to electric power at 100% efficiency, the total power produced would be only about 10% of the total electric power now used.

Figure 10-18 Narrow gorges are ideal sites for hydropower dams from a technical point of view. (Courtesy of the U.S. Department of Energy's American Museum of Science and Energy, operated by Science Applications, Inc.)

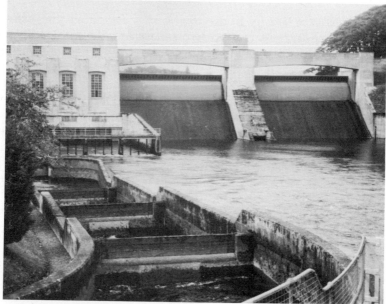

Figure 10-19   Salmon ladder, Pitlochry, Scotland.

The United States is not the only country facing value choices regarding hydropower. Norway has long been known for abundant hydropower. Its slow-melting glaciers have been a source of moving water, which has been channeled down mountainsides to fjords, and in the process has been used to generate electricity, as shown in Figures 10-20 to 10-23.

Some areas of Norway, particularly in the far north, are lacking in sufficient hydropower for further development. The Norwegian government has proposed expanding the hydro facilities by damming rivers such as the Alta. The Alta is a traditionally well-known salmon-fishing river. This could be changed significantly by its adaption as a hydro facility. Furthermore, much of the land that would be flooded by this project is reindeer grazing land (see Figure 10-24). The Lapps have domesticated reindeer and the reindeer are frequently their sole means of livelihood. The overall environmental and sociological consequences are considerable and are currently being debated vigorously.

Figure 10-25 is a schematic drawing of a typical hydro generating facility. The power generated is dependent on the "head" (that is, the height from the surface of the water in the reservoir to the output at the river) and the quantity of water being moved in the actual electrical

Figure 10-20 Glaciers such as this one provide much of the water for Norway's hydro facilities.

Figure 10-21 The water cascading down this mountainside near Bergen, Norway, could provide hydro-electricity.

Figure 10-22 In Norway the water from slowly melting glaciers is funneled down the mountainsides to hydro plants.

Figure 10-23 Hydro power facility near Alvik, in southern Norway, on the Hardanger fjord.

Figure 10-24 Expansion of Norway's hydro facilities is being debated vigorously. Reindeer herds such as this one near Hammerfest traditionally have grazed much of the land that would be flooded.

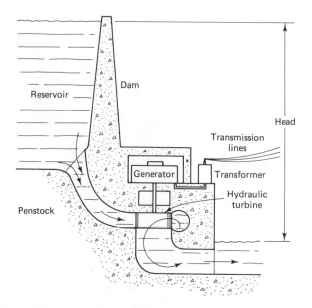

Figure 10-25 Schematic drawing of a typical hydro generating facility.
(Courtesy of the U.S. Department of Energy's American Museum of
Science and Energy, operated by Science Applications, Inc.)

production. The procedure is similar to a fossil fuel plant except that
the turbine is now driven by a moving stream of water. There is some
interest in what is referred to as "low-head" hydropower. These locations involve a series of smaller drops to minimize the environmental
effects.

QUESTIONS

1. Calculate the total electric power that could be generated by hydropower in
    the United States if the gravitational potential energy of every drop of rain
    were converted to power. The total potential energy (P.E.) is given by P.E. =
    mgh, where m is the mass of rainfall on the United States for a year, $g = 9.8$
    newtons per kilogram, and h is the average altitude above sea level of the surface
    of the earth in the United States. To convert to power in watts, divide by the
    number of seconds in a year. Assume 100% efficiency.

2. How much energy is used by a 60-W light bulb in a year if it is turned on for
    2 hours per day on the average? At an electric energy cost of $0.06130 per
    kilowatt-hour, what does it cost to operate the light bulb for the year?

3. Use the information in Appendix C to estimate the total electric energy you
    use for appliances in your house. At $0.06130 per kilowatt-hour what is the
    annual cost?

4. Contact your local electric company and trace the path of the electric power

from the generating facility to your door. At what voltage and current is it generated? Where is the closest substation? At what nominal voltage and current level does the power leave the substation? Where is your customer service transformer? At what nominal current and voltage does it leave this transformer? Where are the voltage regulators in this distribution system?

5. *Load leveling*, that is, attempting to hold constant the electric power demand, can not only reduce the need to use expensive peaking facilities but can also minimize the need to add additional power generation capability. Why is this an important consideration?

6. In this chapter we calculated the ohms of impedance for a 200-W light bulb. The ohms calculated in this way would not equal the ohms measured by an ohmmeter. The computed ohms is for a white-hot filament under normal use. The impedance measured by the ohmmeter is for a cold filament. Which ohm reading is the lesser?

7. Distinguish among electric energy, electric current, voltage, and impedance. Interrelate them mathematically; that is, how would you calculate each?

8. Distinguish between ac and dc power. Why do most power companies transmit ac?

9. Why are appliances such as air conditioners and stoves serviced with 240 V instead of 110 V?

CHAPTER **11**

Nuclear Power: Best Hope or Worst Choice?

Nuclear power is energy that is obtained directly or indirectly from the nuclei of atoms. Atoms that contain intermediate-size nuclei are more stable than atoms with either very large or very small nuclei. The large nuclei, particularly the "heavy" substances such as uranium and plutonium, can be split apart into smaller particles. The "light" substances (hydrogen, helium, lithium) can, on the other hand, be combined to form larger nuclei. In both cases, energy is released. The former process is called *fission* and the latter, *fusion*. Both are considered nuclear energy, but only fission is technologically feasible at present.

FISSION

Fission is the type of nuclear reaction currently utilized in all nuclear power plants, such as the one shown in Figure 11-1. The major fuel is uranium 235 (^{235}U).

Nuclear power has some advantages over fossil fuel power:

1. It does not contribute to changing the temperature of the atmosphere because no carbon dioxide or particulate matter is released.
2. At the present time electric power generated by nuclear energy is less expensive than that generated by fossil plants (Figure 11-2), although those figures may be influenced by government subsidies

Figure 11-1 San Onefre, California, nuclear power plant. (Courtesy of the U.S. Department of Energy's American Museum of Science and Energy, operated by Science Applications, Inc.)

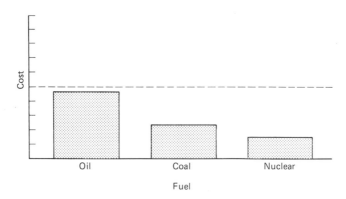

Figure 11-2 Relative costs of power generation. (Courtesy of the Society for the Advancement of Fission Energy.)

to the nuclear power industry, since large amounts of government money have been spent since the 1940s in research and development.

3. In addition, the mining of uranium (mainly in New Mexico and Wyoming), in spite of possible difficulties due to radiation, is much safer than coal mining. For example, in 1982 over 120 U.S. coal miners died, but no one died from nuclear-power-related accidents.

4. Moreover, there is available technology (the breeder reactor) which can be used to extend the supply of nuclear fuel almost indefinitely.

There are, however, a number of serious concerns regarding nuclear power.

1. In spite of the many precautions taken in construction and operation, the possibility of serious accidents or terrorist intervention does exist. No one really knows the actual likelihood of a major problem occurring, for all predictions are simply "best guess" estimates.
2. The operation of nuclear facilities also generates nuclear wastes. The disposal of these wastes is a technological problem and a political nightmare.
3. Although we can estimate the probable effects of various levels of radiation exposure, these are also just estimates. The effects, particularly of very low dosages, are not completely understood.
4. Nuclear facilities create more thermal pollution (Figure 11-3) than that created by fossil fuel plants.
5. Additionally, many economists are worried about rising nuclear construction costs, both the monetary costs and the energy costs.

In spite of the controversy surrounding nuclear power, many

Figure 11-3 Thermal pollution: growth of *Cladophora* along the adjacent shoreline. This increased growth is not known to cause any ecological problems. (Courtesy of the Wisconsin Department of Natural Resources.)

experts believe that coal and nuclear (fission) power are the most likely short-term solutions to our energy dilemma. Alternative sources are either not yet sufficiently developed or too restricted in their applicability to be a short-term solution. It was estimated in 1980 that by 1990 nuclear power would almost double its contribution to U.S. electrical generation.

Certain geographic regions are very dependent on nuclear power for their electrical generation. Overall the United States receives approximately 13% of its electrical energy from nuclear sources. In the early 1980s the states that depended on nuclear power for over 50% of their electricity production were Vermont (78%), Maine (60%), Connecticut (58%), and Nebraska (50%). Figure 11-4 illustrates the regional breakdown of dependence on nuclear power. Figure 11-5 shows the states with operating nuclear power plants. It would take a long time to replace these facilities if it were determined that we should eliminate our nuclear-powered facilities, although several states have seriously considered this option. For example, in 1982 a referendum question permitted the residents of Maine to vote on whether or not to continue to allow fission plants in their state. The result was strongly affirmative.

At least one nation, Sweden, has made the opposite decision. At the present time Sweden, because of its lack of other fuel sources, receives the majority of its electrical power from nuclear energy. However, in the late 1970s the recently elected environmentally oriented government announced that it is determined to phase out the existing nuclear plants. They plan that by the year 2000 these nuclear plants will be phased out and replaced primarily with plants that use peat, a very low grade of coal.

Most European and Asian countries have not taken the antinuclear route. They are expanding their nuclear capabilities. Figure 11-6 reports the number of operating nuclear plants worldwide.

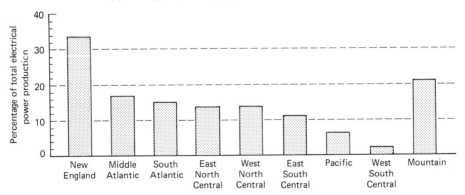

Figure 11-4 Regional dependence on nuclear power. (Courtesy of the Atomic Industrial Forum, Inc.)

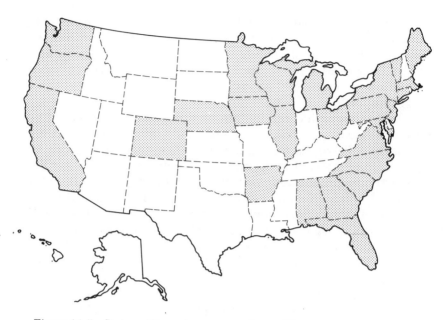

Figure 11-5   States with nuclear power plants. (Courtesy of the Society for the Advancement of Fission Energy.)

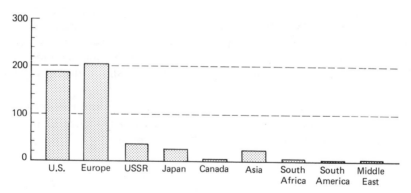

Figure 11-6   Operating nuclear plants worldwide. (Courtesy of the American Physical Society.)

NUCLEAR POWER PRODUCTION

Overall Plant Design

A comparison of the schematic diagrams of nuclear- and fossil-fuel-powered electrical generation facilities shows how they are similar, except for the source of heat (Figure 11-7). In both cases, heat vaporizes water to form steam. The expanding steam drives a turbine, which

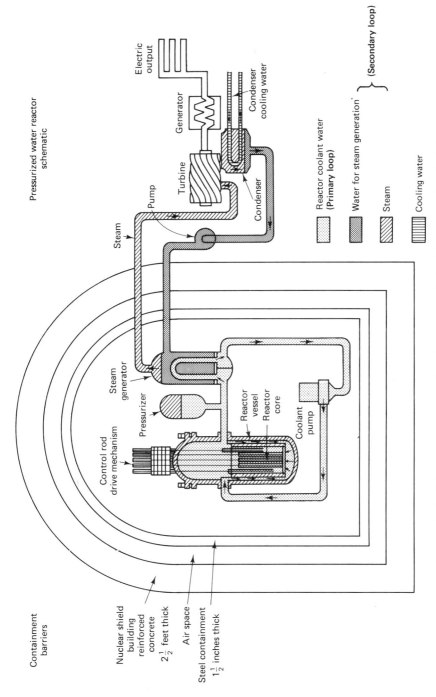

Figure 11-7 Schematic diagram of nuclear-powered electrical generation facility. (Courtesy of the Wisconsin Public Service Corporation.)

causes the generator to produce electric power. The steam is then cooled, recondensed, and recycled.

The fuel for a nuclear plant is generally uranium. Uranium, like many other substances, can be found in several forms. These forms are called *isotopes*. The chemical properties of a substance are determined primarily by the number of protons and electrons found in the neutral atom. All isotopes of uranium have 92 electrons and 92 protons, so the various isotopes of uranium are similar to one another chemically and possess many similar physical properties. However, the isotopes vary in the number of neutrons each possesses. This difference leads to slightly different physical properties. Not all isotopes of uranium are suitable as fuel. Uranium must contain 3 to 4% of the isotope ^{235}U to be suitable as a fuel. ^{235}U is the only isotope that is capable of maintaining a sustained self-propagating reaction (a *chain reaction*) under the conditions found in a typical nuclear reactor. This isotope comprises only 0.7% of natural uranium. This means that the original uranium ore must first be purified and concentrated to about five times its original ^{235}U level. This concentration process is called *enrichment*.

Within the reactor, the neutrons from ^{235}U react with the surrounding medium, producing heat, some fission products, and two to three neutrons, as illustrated in Figure 11-8. The neutrons that start the reaction are present due to the normal radioactivity of uranium. Additional neutrons are, as seen in the equation, produced by the nuclear reaction. The naturally occurring neutrons are not sufficient in number to maintain a sustained chain reaction; normally, too many escape from the edges of the fuel assembly. This leads to the concept of "critical mass." About 1 lb, 0.5 kg, of uranium is the minimum quantity required. Adequate neutrons are then emitted and remain within the bulk of the uranium, available to initiate further reactions.

For a fission reaction to occur, the neutrons must be moving slowly; these are called *thermal neutrons*. Those emitted in the fission reaction are moving rapidly; therefore, they must first be slowed to maintain a sustained reaction. This is accomplished with a moderator, a substance such as water or carbon which surrounds the uranium fuel.

The structure of the fuel assembly in a nuclear plant is critical and significantly different from the fuel found in a fossil fuel plant. The fuel, in the form of uranium dioxide, is compressed into tiny ¼-inch diameter pellets. These pellets are then encased in zirconium or stainless steel tubes called *cladding*. An assembly of these rods comprises the nuclear fuel.

Without a method of control, the fission reactions generated within these fuel rods would rapidly increase and soon be unstoppable. Therefore, rods of cadmium or boron, substances that readily absorb neu-

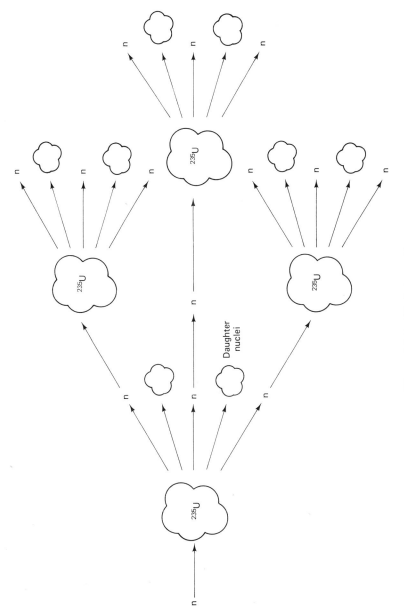

Figure 11-8 Fission reaction.

trons, are interspersed among the fuel rods. These *control rods* (Figure 11-9) can be raised or lowered to regulate the rate of the reaction of neutrons with uranium. Figure 11-10 illustrates a typical overall nuclear facility.

Figure 11-9 Nuclear fuel assembly control rods.

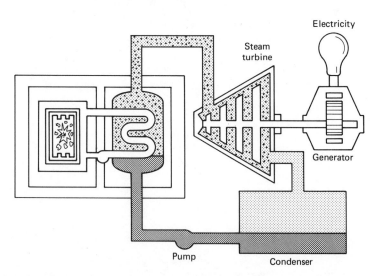

Figure 11-10 Overall schematic diagram typical of a nuclear facility. (Courtesy of the U.S. Department of Energy's American Museum of Science and Energy, operated by Science Applications, Inc.)

Safety Features

There are a number of safety systems in a nuclear facility designed to eliminate the dangers of nuclear accidents. The first safety feature is the containment building, which encloses the actual reaction vessel (Figure 11-11). This is constructed of reinforced concrete and is dome shaped to maximize its strength against explosions from within and without. The nuclear reactor itself cannot explode for there is insufficient ^{235}U present to do so. Sabotage is, however, a continual concern. The containment shell is designed primarily to contain high-pressure steam contaminated with radioactive particles that might occur during malfunction. The reactor itself is underground, surrounded by water and more concrete. Also in the containment building are the steam generators. This steam is used to power the turbines to generate electricity, as shown in Figure 11-12 and discussed in Chapter 10.

This containment area is sealed from the other sections of the facility to assure that all radioactive releases are restricted to this area (see Figure 11-13). Anyone who enters must wear special clothing, overalls, boots, cotton gloves, and a hair net. When leaving the area, these clothes are removed. The boots and overalls are washed; the gloves and hair net are disposable. Special areas are even marked on the floor. When removing one's boots, for example, first one boot is removed, then the other foot can be put outside the restricted area; when the second boot is removed, that foot can also be placed on the outside.

One of the major concerns regarding nuclear power plants involves what is called a *loss of cooling* (water) *accident*. This is what happened at Three Mile Island in 1979. The loss of the water around the fuel rods stops the fission reaction, for the water serves as the moderator which is necessary to slow the neutrons sufficiently for the reaction, but the fuel rods continue to increase in temperature, due primarily to the radioactive decay of the fission products that have been created. If this continues, the fuel rods can reach such a high temperature that the cladding melts and the radioactive fuel can collect in a puddle on the cement floor. This hot fuel theoretically could melt through the cement floor and escape to the environment. This could have exceedingly serious environmental consequences if it got into the natural water systems. A meltdown would be a very expensive nuisance, but as long as the hot fuel did not melt through the floor, it would not be particularly dangerous even to the operating crew. If, at worst, the shell should crack and let radioactive contaminated steam escape, the steam will rise up into the atmosphere and drift downwind. Those downwind may have to avoid the radioactive "rain," and the easiest way to avoid it is to move away for a few days until the mess can be cleaned up. Recent independent investigations, such as by the Kemeny Commission after

(a)

(b)

Figure 11-11 (a) Construction of a reactor vessel. (Courtesy of the Tennessee Valley Authority.) (b) Construction of the Commonwealth Edison Braidwood Nuclear Station containment building. (Courtesy of Commonwealth Edison.)

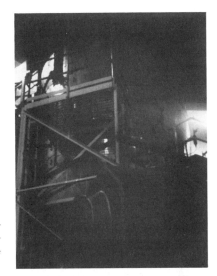

Figure 11-12 Turbine and generator systems in a nuclear plant. (Courtesy of the Wisconsin Public Service Corporation.)

Figure 11-13 Door seals such as these assure that radioactivity is contained. Note also the clothing of the operator.

the Three Mile Island affair, have indicated that these meltdown accidents are probably more likely than once thought, but the likelihood that the radioactivity would escape the containment building is almost zero.

To minimize the possibility of occurrence of a loss of cooling accident, modern reactor facilities have emergency water supply systems such as sprinklers on the ceiling of the containment building and pumps (Figure 11-14) to supply additional water to the reactor vessel. In case of an accident, the whole containment building can be filled with water. As is perhaps obvious, the installation of these safety features, although necessary, does greatly increase the capital costs of nuclear plants. Since the Three Mile Island incident in 1979, the added safety requirements have increased the construction costs of new plants and have led to cost overruns for those facilities under construction. It has also delayed both the start of new plants and the completion of on-going projects. These added expenses eventually get passed on to the consumer.

Radioactive Waste Disposal

The fuel in a reactor must be replaced periodically. Typically, one-third of the fuel is replaced each year. The new fuel is not particularly dangerous, but the old fuel, containing the fission products, presents a major hazard. Currently, these spent fuel rods are stored on the site of the nuclear generating facility (see Figure 11-15). About 0.5 to 1.0 cubic yard of such material is generated per nuclear reactor each year.

Several types of disposal sites for highly radioactive waste are being considered. Burial in abandoned salt mines is one possibility, and another is burial in shafts dug 1000 ft deep into granite bedrock (Figure 11-16). Other remote possibilities include sending the waste in a rocket

Figure 11-14 Pumps that supply emergency cooling water.

Figure 11-15 Spent fuel rods are stored on-site, under water, in locations such as these.

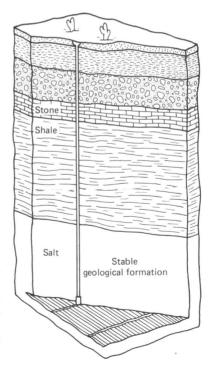

Figure 11-16 Proposed nuclear waste disposal methods. High-level wastes might be stored deep underground in stable geologic formations such as granite or salt deposits. (Courtesy of the Society for the Advancement of Fission Energy.)

toward the sun or burying it in Antarctic glaciers, on the seafloor, or in deep ocean trenches.

The technology required for some disposal techniques has been perfected, but no sites have yet been established. Low-level wastes (including the disposable hair nets and gloves) have been disposed of in sites such as Barnswell, South Carolina. The primary problems associated with waste disposal are political. A complicating factor is the government requirement that all wastes be kept available for retrieval during the next 100 years. The reasons for the requirement is the possibility of recycling these materials. If recycling facilities are developed, this spent fuel could be an exceedingly valuable resource.

Breeder Reactors

One of the potential limitations of nuclear power is the lack of uranium fuel. Our estimated reserves will last less than 50 years. These resources could be appreciably extended by use of a *breeder reactor*. The basic principle is simple. The fuel rods are surrounded by natural uranium or even "waste" material from the spent fuel rods. Some of the neutrons emitted in the fuel rods are captured by this surrounding material, converting it to plutonium, which can also be used as a fissionable nuclear fuel.

There are some potential problems associated with breeder reactors. First, they are not as stable as conventional reactors, hence accidents could occur more readily. Second, the fuel generated is high-grade plutonium, which can be used to manufacture nuclear weapons. This might lead to more nuclear proliferation. Third, plutonium is also an exceedingly dangerous carcinogenic (cancer-causing) substance.

The United States has no breeder reactors at present. In the mid-1970s, the federal administration cut the funds for the Clinch River project, our pilot breeder reactor project. A number of other countries, however, including France, Britain, and USSR, do have operating breeder reactor facilities.

RADIATION HAZARDS—IN PERSPECTIVE

Many people are seriously concerned about the potential radiation hazards from nuclear power plants. How serious are these dangers? How much harm have nuclear plants already done? How much harm are they likely to do in the future, as the number of plants increases? To answer these questions, it is necessary to look closely at the nature, major sources, and the levels of nuclear radiation.

Radiation is a very general term. Electromagnetic radiation, described in Chapter 3, is one type of radiation that can be produced as a result of changes within the nuclei of atoms. Nuclear radiation also includes high-energy particles that are emitted as a result of these nuclear reactions. These particles include alpha particles (the nuclei of helium atoms) and beta particles (electrons). A major source of nuclear radiation is the radioactive decay of the fission products from a nuclear power plant.

This type of radiation is called *ionizing radiation*. It can detrimentally affect living tissue by altering its molecular structure. The conventional unit for the measurement of energy retained by living tissue is the radiation absorbed dose, or *rad*. Since different types of nuclear radiation vary dramatically in their penetrating power and the resultant amount of damage they can cause, the rad must be weighted for differing biological effects. The weighted measure is the *rem*. In contrast to the rad, it is independent of the type of radiation. When discussing low-level radiation, dosages are typically expressed in millirem (mrem), which is equal to 0.001 of a rem.

Nuclear power plants are not the only source or even the major source of nuclear radiation. Even prior to the development of nuclear power, every animal and every plant that has ever lived on this earth had been continuously subjected to nuclear radiation. Both cosmic rays from space and radiation from radioactive trace elements in soils provide the average U.S. resident with 35 mrem per year of nuclear radiation exposure. Radioactive potassium 40 and carbon 14 present in everyone's body contribute another 22 to 27 mrem per year. Medical diagnostics such as X rays or radiopharmaceuticals contribute, on the average, another 70 mrem per year. Nuclear weapons test fallout and all energy production and use add another 4.4 and 3 mrem per year, respectively. The average U.S. citizen is therefore exposed to about 100 mrem per year, depending on location and occupation.

Figure 11-17 illustrates the average natural background radiation that can be expected in each state. Notice the contrast between high-altitude Colorado, with its large granite deposits containing naturally occurring uranium, thorium, and their daughter elements (the fission products) and Florida, with a very low average altitude and primarily limestone rock.

One's occupation or life-style also affects the background radiation levels experienced. Activities such as engaging in a large amount of air travel can increase the background exposure levels for some people to 200 to 300 mrem per year.

How do these levels correspond to radiation doses considered hazardous to health? Radiation dosage levels can be classified as high level (above 5000 mrem) or low level (below 5000 mrem). About 500,000 mrem is considered to be a lethal dose to 50% of the human beings who

Figure 11-17 Average natural background radiation, by state (cosmic, terrestrial, and internal body radiation in millirems per person per year). (Courtesy of the National Society of Professional Engineers.)

experience it; 50,000 mrem is the level where the first detectable physiological impacts can be seen; 5000 mrem is the maximum allowable average annual exposure to industrial workers.

Certainly, the background levels we have been discussing fall well below these levels. However, do relatively low exposure levels also affect human health? The answer may be "yes." Radiation affects a person's living tissues by altering the molecular structure of cells. It typically can lead to cancer or genetic defects. The actual danger of very low level exposures is, however, unclear. In general, the elderly and the very young (particularly, the *in utero* fetus) experience the greatest risk. Studies have generally shown that the size of the dose is not directly proportionately to the response or intensity of the effect. In a few cases, exposures as low as 10 rem (10,000 mrem) have been thought to result in the birth of mentally defective children; generally, however, there are no consistent findings of defects at doses between 1 and 20 rem.[1]

To estimate the number of cancer fatalities caused by ionizing radiation, one cancer fatality has been assumed to occur per 5000 person-rem. The person-rem is a measure of population exposure rather than of individual exposure. It is the added doses of each of the exposed per-

[1] Leonard Sagan, "Radiation and Human Health," *EPRI Journal*, September 1979.

TABLE 11-1   Predicted Cancer Fatalities due to Ionizing Radiation, General U.S. Population: Average Dose*

	mrem/year	Radiation Fatalities Total Number of Persons in U.S. per Year	Radiation Fatalities Per Million Persons per Year
Medical-diagnostic	70	3080	14
Cosmic radiation	35	1540	7
Terrestrial (rocks, soil, etc.)	35	1540	7
Potassium 40 in food	20	880	4
Nuclear weapons fallout	4.4	194	0.9
Use of natural gas in homes	2	89	0.4
Burning of coal	1	44	0.2
Sleeping with another person	0.1	4.4	0.02
Consumer products (TV, etc.)	0.03	—	
Total	168	7377	

*Assume one cancer fatality per 5000 person-rem.

Source: Courtesy of the National Society of Professional Engineers.

sons and assumes a linear relationship between dose and effect. For example, 100 persons exposed to 5 rem each would be 500 person-rem. The ultimate statistical effect is assumed to be the same as if only one person received 500 rem.

Table 11-1 summarizes data compiled for the National Society of Professional Engineers regarding predicted cancer fatalities in the general U.S. population. Table 11-2 considers special population groups. A normally functioning nuclear plant discharges no more radioactivity than a comparably sized coal-fired plant, due to the radioactive isotopes found in coal. Three Mile Island is the most serious nuclear power plant incident to occur to date in the United States. It is worth noting that for the average person located within 50 miles of the plant the increased dose was 1.5 mrem. The maximum dose to a person immediately off-site was 83 mrem. These added exposures should not create any measurable health hazard.

THERMAL POLLUTION

All electric generating facilities and many industries create *thermal pollution*, the emission of excess heat into the environment. The problem is, however, particularly severe with nuclear power plants. It is necessary to maintain the temperature inside a nuclear reactor somewhat

TABLE 11-2   Predicted Cancer Fatalities due to Ionizing Radiation:
Special Groups in the United States*

	mrem/Year	Radiation Fatalities per Million Persons per Year
Transcontinental jet crew (70 hours of flight time per month)	385	77
Nuclear power occupational (average of all monitored employees, 1973–1977)	365	73
Western representative campaigning for Senate (extra air travel)	200	40
Grand Central Station, Vanderbilt St. entrance, 40 hours/week	120	24
Increased natural background in Colorado compared to Florida	80	16
Being an average representative: 50-hour workweek, 25 trips home by air per year	20 50 ─── 70	14
Population within 50-mile radius of Three Mile Island: 3/18/79–4/7/79 (average per capita dose for 2.1 million persons)	1.5 mrem total	0.3
Nuclear power plant guard	1	0.2

*Assume one cancer fatality per 5000 person-rem.

Source: Compiled by Mike McCormack. Courtesy of the National Society of Professional Engineers.

lower than inside a fossil fuel combustion chamber in order to reduce damage to the fuel and fuel rods. This lower temperature decreases the efficiency of the electrical generation in accord with the second law of thermodynamics, and thus increases the amount of waste heat emitted. Typically, 60 to 70% of the energy originally present in the fuel forms thermal pollution at the plant site. The cooling water released by nuclear power plants is about 10°C above its intake temperature. Note that in a nuclear plant the water in contact with the fuel is not that which is discharged. The heat from this steam is extracted by an indirect (noncontact) heat exchanger; in the process a second stream of water is converted to steam. This stream is used to operate the turbine. It is cooled by a third water stream in a (noncontact) condenser and it is this water that is discharged. This water has never been near the reactor and is radioactive only to a slight degree, caused by secondary emission processes. The other two water streams are never discharged during normal operation.

When the warm water is discharged, it creates a plume of warmish water downstream from the point of discharge. Fish who like warm water will move into the plume, those who do not will move out. The 10°C variation is less than the normal climatic change of temperature of a stream or bay, so fish have little difficulty adapting. Marine life that does not move around is altered depending on the prevailing temperature. Warm water has one type of population, cold water a different.

To date, no harmful biological changes have been noted due to the release of warm water from any power plant. Fossil fuel plants have been discharging their warm water into rivers and streams for over a century without causing any noticeable harmful change in the aquatic population. The Connecticut River below the Haddan Neck nuclear plant, for example, has been biologically monitored since the early 1960s, and no deleterious changes have been noted. In the Great Lakes, fishermen tend to concentrate their efforts in the discharge plumes of nuclear power plants, for it is there that the desirable fish concentrate.

FUSION REACTORS

Research is continuing into fusion power. Fusion is a duplication of the types of nuclear reactions occurring in the sun. For these reactions to occur on Earth, several practical problems must be solved. It is necessary to reach a high enough temperature for a long period of time to have the substances react. This is exceedingly difficult, for temperatures of 100 million degrees Celsius are required. No conventional material can withstand these temperatures directly.

Two proposed technological solutions are under investigation to solve the confinement problem: (1) magnetic confinement of the hot plasma (Figure 11-18), or (2) laser-induced fusion (Figure 11-19). In

Figure 11-18 Magnetic confinement equipment, intended to initiate controlled nuclear fusion. This is the Tokamak reactor, located at Princeton University. (Courtesy of the U.S. Department of Energy's American Museum of Science and Energy, operated by Science Applications, Inc.)

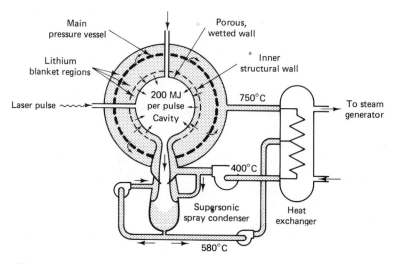

Figure 11-19 Schematic of laser-induced fusion. (Courtesy of the U.S. Department of Energy's American Museum of Science and Energy, operated by Science Applications, Inc.)

the latter, a solid fuel pellet of lithium is impacted by a very high energy laser beam. The laser light compresses the lithium into a more dense state. There is a very short confinement time, but very high temperatures can be reached. Magnetic confinement, on the other hand, involves the use of very intense magnetic fields to confine a hot sample of a gaseous fuel. The relatively long confinement time minimizes the actual reaction temperature required. The major problem associated with both of these methods is the inability to obtain net energy from the process.

If fusion were possible, it could be a tremendous breakthrough. Typical reactions and the energy releases are

$$d + d \longrightarrow t + p + 4 \text{ MeV}$$

$$d + d \longrightarrow {}^4\text{He} + n + 17.6 \text{ MeV}$$

where   d = deuterium (hydrogen with an "extra" neutron in the nucleus)

   t = tritium (hydrogen with two "extra" neutrons in the nucleus)

   n = neutron

   ^{4}He = helium

   MeV = million electron volts = 1.5×10^{-16} Btu

QUESTIONS

1. Describe the arguments for and against nuclear power. Which ones are most persuasive to you? Why?

2. Why is it impossible to phase out quickly our dependence on nuclear power, even if it were determined desirable to do so?

3. Compare and contrast nuclear- and fossil-fuel-fired electrical power generation facilities.

4. Describe how a nuclear fission chain reaction is established. What initiates it? How is it propagated? How is it kept under control?

5. Describe typical safety features at a nuclear power facility. What type of safety problem(s) is each designed to minimize?

6. What methods are currently under investigation for the potential disposal of high-level radioactive wastes? What would be an advantage of each method? A disadvantage?

7. Compare the background radiation that people living in Colorado experience compared to those in Florida. What causes the major difference? How does your state compare to these two extremes? What major geological features account for this background radiation level?

8. What are the major causes of additional radiation exposure which are job dependent? That is, why are some jobs more likely to lead to increased exposure?

9. Describe the operation of a breeder reactor. What factors have led the United States to be very cautious in the implementation of this technology?

10. Describe typical fusion reactions. What are the major difficulties preventing its commercial development?

CHAPTER 12

Solar Energy: Harnessing the Sun

The celestial neighbors that share the solar system with the earth interact with the earth in two ways. First, the gravitational forces exerted most predominately by the moon and the sun maintain the earth in its orbit about the sun. The only direct impact of extraterrestrial gravitation that is observable is the tide. The other interaction from outer space is the energy we receive from the sun. The significance of solar energy as a global phenomenon was introduced in Chapter 3. The intent of this chapter is to examine the technologies that people have developed to use the sun's energy to meet personal needs.

One of the first questions that must be asked when considering the development of a technology appropriate to an energy source is whether there is enough energy available from that source to make the proposed venture a worthwhile effort. From this standpoint, the sun has some impressive credentials. The rate of solar energy input into the upper atmosphere of the earth is about 180,000 trillion watts (or 1.8×10^{17} W). The current rate of energy use by the people of the earth is about 1 trillion watts. This is equivalent to saying that one day's worth of solar energy is equivalent to the total energy used of the earth in 500 years. The total energy received by the earth in 3 days is equivalent to all the presently known reserves of coal, oil, and natural gas in the world.

Solar energy is abundant, but on the earth's surface it is not evenly distributed at one location from day to day due to climatic factors or with the same concentration from place to place due to the curvature of the earth. Relative solar energy distribution in the United States is illustrated in Figure 12-1. It is interesting to note that in many instances it is economical to invest in solar energy technology in geographic regions that are less than ideal in solar input. The energy required for space heating in these regions may be high, so that a wide range of complementary energy alternatives that reduce fossil fuel consumption are economically attractive.

Solar energy technology can be divided into two major categories. There are large solar systems, where the energy derived is collected and distributed in a fashion similar to conventional energy sources that are classified as public utilities. Small solar systems are those where all the energy is used in a single building, which may be a home or a commercial or industrial building.

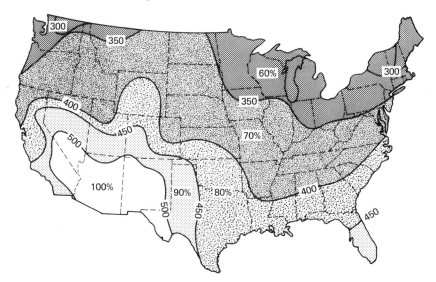

Figure 12-1 Relative distribution of solar energy in the United States. (Courtesy of the U.S. Department of Energy's American Museum of Science and Energy, operated by Science Applications, Inc.)

LARGE-SCALE SOLAR ENERGY PROJECTS

There are several solar energy systems that can be considered as large-scale projects. One of the most revolutionary is a solar satellite. The idea of a solar satellite, which is described in Figure 12-2, was first proposed in the late 1960s. For the past 10 years the technology has been available, but economic and environmental considerations have

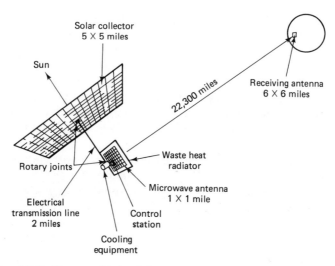

Figure 12-2 Energy obtained from a solar satellite. (Courtesy of the U.S. Department of Energy's American Museum of Science and Energy, operated by Science Applications, Inc.)

delayed mounting an all-out national effort. As illustrated in Figure 12-2, this satellite would cover perhaps as much as 25 square miles, about 70% of a township, and then send the collected energy back to the earth in the form of microwaves. The satellite would be located at a distance of 22,300 miles above the earth so that its orbiting period is 24 hours, which in effect "parks" the satellite over one specific location above the earth. A 36-square-mile microwave target, or one township, on the earth would be designated for receiving the energy. This region would not be an appropriate place for human habitation due to a high concentration of microwaves coming in from the satellite.

There are some problems associated with the solar satellite system. One obvious consideration is cost. Estimates of satellite cost range from a low of $12 billion to as much as $60 billion. The range in estimates is large because the cost of each satellite depends to a great extent on how many will ultimately be built. It will be necessary to construct the satellite in space since it is far too large and too fragile to launch from the earth. The "home in space" for the construction workers will cost at least $200 billion. Some of the most recent cost estimates state that household electricity originating from the satellite will produce electric bills of about $600 per month.

Another problem of a solar satellite is the environmental impact of the satellite target on earth. What happens when part of the earth is continuously bombarded with microwaves? A complete answer is unknown, but reasonable speculation can lead to the prediction that it is unlikely to raise real estate values. The solar satellite would also be an appealing military target for an unfriendly nation. If the satellite were captured, the power source might also become available to the "other side" or a new target could be selected, such as a large metropolitan area. In 1981 the federal government indicated that interest in this project was beginning to wane.

Another large technology design is the use of several hundred solar collectors that concentrate solar energy, which is in turn converted to electric power by conventional means. Sunlight strikes a relatively large area of mirrors on Earth, which is reflected to a central receiver, often called a *power tower*. There water is heated to a high temperature. There is a boiler mounted on the tower which converts the water to steam. The steam drives a conventional electric generator. The power tower would probably produce about 10% as much electricity as a large nuclear power plant. Twenty-five thousand mirrors would cover a 400-acre area. A 10-megawatt plant is currently operating in Barstow, California, and is producing electricity at a cost of about $2.00 per kilowatt-hour. This compares with electricity costing $0.05 to $0.15 per kilowatt-hour from fossil fuel plants. Estimates are that if 15,000 square miles are covered with mirrors, about 50% of U.S. electrical needs can be met. This land can also be used for other purposes as is illustrated in Figure 12-3.

Figure 12-3   Solar "farms" where land is used for collecting solar energy and range land for cattle grazing. (Courtesy of the U.S. Department of Energy's American Museum of Science and Energy, operated by Science Applications, Inc.)

If solar energy were used to produce electricity and feed into the existing power grid, estimates of the amount of the electricity that can be produced range from 1 to 8 percent of U.S. needs. One percent of U.S. electricity needs from sunlight will require an investment of $10 to $20 billion.

Perhaps as much as 20% of the nation's total energy might be available from sunlight if extensive efforts are made in the area of residential solar heating. Solar space heating is the primary method of using energy in a small-scale energy project.

SMALL-SCALE SOLAR ENERGY PROJECTS

For centuries people have used solar energy directly to dry food, dry clothes, and grow plants in sheltered environments. Small solar energy projects can be classified as either active or passive solar systems. The most obvious component of an active solar system is the solar collector, which is usually attached to the roof of a building. The amount of energy available to a collector averages 3770 calories per square meter each day, which is equivalent to 58 Btu per square foot each hour depending on the cloud cover and angle of incidence of the sunlight on the collector. In July the value increases to a range of 70 to 115 Btu per square foot each hour. A furnace in an average-size home may be able to provide 100,000 Btu each hour if it ran continuously. Fortunately for the consumer, it is not necessary for a furnace to run continuously because using the cheapest form of fossil fuel, natural gas at $0.50 for each 100,000 Btu, this continuously running furnace would cost the homeowner $360 a month to operate.

Active Solar Systems

Collectors consist of a box with one or more panes of glass as a cover (see Figure 12-4). The glass plates, in addition to allowing sunlight to enter the collector, serve as insulating material to keep the heat inside the collector. Visible light travels through glass but most of the heat does not. Using more glass panes with dead air space between them causes less heat to escape. Increasing amounts of glass increase the amount of solar energy absorbed and reflected by the glass itself, which reduces the effectiveness of the collector. Usually, one to three panes of glass are appropriate.

Heat is removed from the box by heating a fluid that flows through tubing that crisscrosses the box. The fluid may be either a liquid such as water or a gas such as air. Water is more efficient than air in transporting heat from the collector, but water leaks can cause significant problems, and water without antifreeze freezes in the winter.

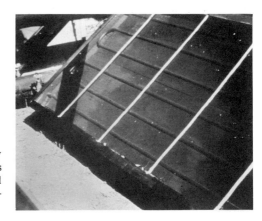

Figure 12-4 Solar panel. (Courtesy of the U.S. Department of Energy's American Museum of Science and Energy, operated by Science Applications, Inc.)

There are several factors to consider in comparing solar collectors. The cost per square foot of collector surface is one direct comparison, but the relative efficiencies of the collectors must also be considered. A poorly insulated cheap collector may not be a bargain. If a cheap collector that costs one-half as much per square foot as a more expensive collector is only 50% as efficient as the more expensive collector, the consumer does not save any money buying the cheaper system because he or she will have to purchase two cheap collectors to obtain the same amount of energy provided by the more expensive one. The efficiency of solar collectors decreases as the outside temperature goes down because the rate of heat flow from the inside of the collector to the outside depends on the difference between inside and outside temperatures. When the temperature is very cold (below $0°$ F) a poorly insulated solar collector may lose so much heat to the outside that almost no heat is transmitted to the conducting fluid, and consequently, the heating efficiency of the collector may be nearly zero.

Another point to consider when purchasing a solar collector is the difference between the highest temperature the solar collector can attain in the summer, when perhaps there is no fluid flowing in the system, and the melting temperature of some of the components inside the collector. A poorly designed collector may become so hot in the summer that some plastic parts may melt. Replacing parts in a collector mounted on a roof may be expensive or dangerous or both.

The type of heat storage system required for storing the energy collected by the solar collector depends on what the energy is to be used for and the kind of fluid in the collector. If the energy is used to assist the household water heating system, the fluid is usually water. The water heated in the solar collector is in a closed loop with the other end of the loop extending into a heat exchanger. Household water that has been heated through the heat exchanger is usually stored in a large holding tank of perhaps 150 gallons. Usually, 1.5 gallons of water can be stored for 2 or 3 days for each square foot of collectors. Water from the storage tank can subsequently be sent through the conventionally fueled hot water heating system to increase the temperature to the desired level if necessary.

If the solar heating system is used primarily for space heating, the fluid may be air. Rocks often serve as the heat storage system and 60 lb of rocks can accommodate the energy from each square foot of the collector.

Other storage systems involve using the heat from the collector to convert a solid salt to a liquid. Then when solar energy is not available from the collector, the salt reverts back to a solid releasing a substantial amount of heat in the process.

Active solar heating systems that are practical in terms of size and economics may provide from 20 to 70% of the heat energy needed for a home. The cost of the solar system is definitely an add-on cost because some form of conventional heating will still be required. Prices of solar systems vary, but it is not unreasonable to expect that an active solar system that provides 50% of the yearly heat energy for a home may increase the cost of the home by 10 to 20%. The addition of $12,000 to a mortgage at 12% interest means an interest payment of more than $1400 per year. At the present time $1400 spent on fossil fuel will cover and in most cases exceed 50% of the energy costs of a typical household. It may also be necessary to replace substantial parts of the system during the lifetime of the house.

Passive Solar Systems

An alternative to the active solar system is a passive system. In a passive system the building serves as the collector and storage unit. There are several attractive features to a passive system. Since modifications are in the basic design, a passive solar building is an alternative to and not a modification of a conventional structure, meaning that there are fewer add-on costs. A passive solar building can attain heating efficiencies that are near those obtained using active systems, and many people prefer the passive solar homes to conventional homes.

There are three basic features that all passive solar buildings have in common. First, windows are strategically placed. In the northern hemisphere a south-facing window is an energy "gainer"; that is, the amount of solar energy that enters the window during the daylight hours from the outside exceeds the amount of heat energy that escapes from inside through the window after the sun sets. North-facing windows are always energy "losers." A passive solar building then concentrates windows on the south side and has very few, if any, windows on the north side. Window coverings, such as insulated drapery or Styro-foam panels, may also be used to reduce heat loss at night.

Overhangs are placed above the windows to allow sunlight to enter during the winter when the sun is nearer the horizon, but during the summer when the sun is nearly vertical the overhang blocks the sun's rays from the window. It is also useful to plant trees that loose their leaves in front of windows on the south side of the house. In the summer these trees shade the windows but are not an obstruction to sunlight in the winter.

Passive solar buildings depend on a large mass of material to absorb solar energy during the daylight hours and then slowly radiate this energy at night. This massive material may be a stone or brick wall or a floor that is in the path of the sun light entering the windows. It may also consist of water that is stored in barrels or other containers. These heat-absorbing masses also reduce the temperature ranges that are associated with large window areas on the south side of a house. (See Figures 12-5, 12-6, and 12-7.) Interior wall openings and possibly small fans are arranged in an effort to assist natural convection currents that transport the heat into the remainder of the house.

There are numerous advantages to passive solar heating, including low construction cost with familiar materials, reliability because there is no mechanical apparatus, enjoyable living environments due to exten-

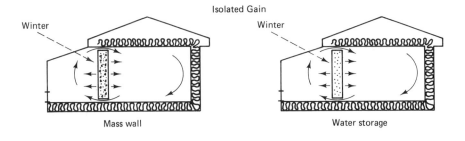

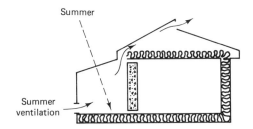

Figure 12-5 Isolated gain: Solar radiation is collected in a space separate from the living space. This space may be used as a greenhouse.

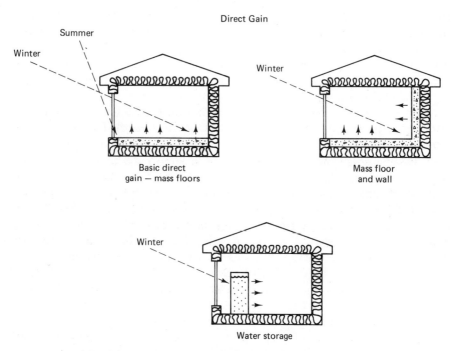

Figure 12-6   Direct gain: Solar radiation travels through and is collected in the living space.

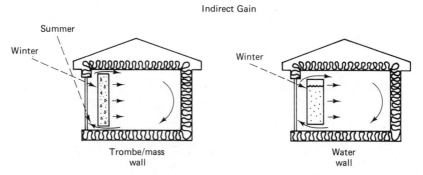

Figure 12-7   Indirect gain: Solar radiation does not travel through the living space. This minimizes large temperature changes. Two variations for indirect gain buildings include the mass storage wall (known as a trombe wall) and the water storage wall. Both systems may or may not have vents to allow natural circulation of the air between the glazing and the mass. The vents allow more heat to transfer to the room air by convection. Proper shading and night insulation are important to control heat gain and loss from the building.

sive sunlight in the home and an aesthetically appealing environment, and a performance efficiency that can be nearly equal to that of active solar systems.

Photovoltaics

Photovoltaic cells provide for the direct conversion of solar energy into electrical energy. This is accomplished by using solar energy to remove electrons from atoms that are located on a crystal. If these free electrons can be encouraged to flow through an electric wire, an electric current is established. The position the electron occupied before its encounter with the solar energy is called a *hole*. The hole is available for occupancy by another electron. The result is that solar energy striking an appropriate surface causes electrons to move into different energy regions. The difference in energy between the electrons in the two regions is the energy that is utilized by an electrical appliance.

An individual solar cell is capable of producing a current of 0.030 A per square centimeter and a maximum potential of 0.500 V; however, these maximum valves are achieved only under very different load conditions. Sunlight under the most ideal conditions delivers about 1000 W of energy for a square meter of area. A more realistic number to work with, however, is 800 W per square meter. A photovoltaic cell with a collecting surface of 1 m^2 would produce about 150 W of power at about 15% efficiency (see Figure 12-8). Maximum efficiency is probably between 20 and 25% using silicon materials because sunlight has a large amount of radiation that is not an appropriate energy level to cause electron activity in a photovoltaic cell.

The major problem associated with photovoltaic systems is the price. From 1976 to 1980 the cost per watt had decreased by a factor

Figure 12-8 A large bank of photo-voltaic cells. (Courtesy of Stone and Webster.)

of 10. Projections to 1990 suggest that the cost at that time may be 1/100 of the 1976 cost. At that time photovoltaic systems would be highly competitive with conventional energy sources. At the present time it would cost the average household more than $100,000 to install sufficient photovoltaic cells to meet all the needs of their home from this source.

For small energy systems such as households, photovoltaics may someday be a very practical method of obtaining electricity. Excess electricity could be sold to power companies to enable them to better meet their peak-load requirements. However, the production of large amounts of electricity will require large expanses of land. Several million acres of land would be needed to produce less than 10% of U.S. electrical energy needs. Given the current efficiency of photovoltaics, this would require a substantial commitment of dollars and space.

QUESTIONS

1. Why might it be more practical to build a solar energy collector in Minnesota than in Arizona when Minnesota receives only about 60% of the solar energy that Arizona receives?

2. How can a satellite be described as "parked" above a certain location on the earth's surface?

3. Is there any way that a solar satellite could alleviate or contribute to the atmospheric problems described in Chapter 3?

4. What is the advantage of having a solar panel on a movable platform so that over the course of a day the solar panel can "follow" the sun?

5. If two pieces of glass in a solar panel do a better job of insulating the panel to prevent heat from escaping than one piece of glass will, why not use four or six or eight pieces of glass for even better insulation?

6. Why is the collector that costs the least per square foot of surface not necessarily the best bargain? What are some other important considerations?

7. How can a solar collector become less effective as temperature goes down on a sunny day?

8. If passive solar heating were to be implemented on a large scale, it is possible to imagine a high percentage of new single-family homes and perhaps duplexes using passive heating. There are several reasons for not expecting large savings in total energy used for home heating in the near future. One reason is that the average life for a home is more than 40 years. Why is this a reason, and what is another reason?

9. If the maximum efficiency of producing electricity from a photovoltaic cell is 25%, how does this compare with the efficiency of producing electricity by other methods?

CHAPTER 13

Wind Power:
Is the Breeze Free?

ORIGINS AND CHARACTERISTICS OF WINDS

Wind energy actually originates from solar energy interacting with the earth's surface and atmosphere. Wind patterns can be divided into three major categories: global, regional, and local.

Global Wind Patterns

Global wind patterns (Figure 13-1) consist of three major air circulation cells. Air masses moving into the equatorial regions from north and south of the equator are warmed because they are traveling close to a warm surface that has received a larger than average share of sunlight. When these air masses meet near the equator, they rise. In the upper atmosphere, as some of the air travels north away from the equator, it cools and becomes more dense. At about 30° latitude north of the equator, this air mass meets the upper atmospheric air from the next cell. The heavy dense air descends back to the earth's surface, where it again moves either north or south.

The regions near the equator where the air ascends are called low-pressure regions. Since the air is cooling, the relative humidity rises and precipitation often occurs. Therefore, the vegetation on the landmasses near the equator is typically dense tropical rain forests. Air descends in a high-pressure region. Air under these conditions is increasing in tem-

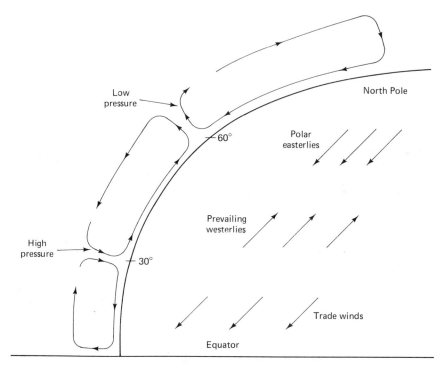

Figure 13-1 Major air circulation cells in the northern hemisphere.

perature, with less relative humidity and precipitation unlikely. Many of the notable deserts of the world, such as the Sahara and the Mojave, are located near 30° latitude.

A cell similar to the one in the tropics exists in the polar regions, creating a high-pressure region at the poles and a low-pressure region near 60° latitude. The resulting global wind patterns one would expect near the earth's surface are a wind from the north between 30° and the equator, a south-to-north wind between 30 and 60° and a north-to-south wind between the poles and 60°. The rotation of the earth, however, modifies the wind directions so that the winds near the equator, called the trade winds, travel from northeast or southwest, the middle latitude winds generally move from west to east, and the polar winds are called the polar easterlies.

Regional Wind Patterns

Regional weather systems, which often move across a continent in the direction of the winds associated with the global cells, significantly alter the wind directions in any given locality from day to day. When low-pressure systems move from west to east, in the middle latitudes, air moves toward the center of the system in a spiral path in a counter-clockwise direction. Air moves away from a high-pressure system in a clockwise direction.

Local Wind Patterns

Local geographic features can also have a significant impact on wind direction and intensity. During the day, large bodies of water are often colder than nearby land masses. Consequently, as warm light air above the land rises, it is replaced by cooler air blowing from the water body to the land. Because the land cools off more rapidly at night than the water, the wind direction will reverse and a land breeze results.

Winds and wind direction are the direct result of uneven heating of the earth by the sun. One cause of the uneven heating is due to the curvature of the earth, which causes sunlight to strike at different angles with resulting differences in the intensity of sunlight. A second reason for uneven heating is due to the difference in the amount needed to produce a specific temperature change and the fact that some surfaces reflect sunlight better than others. The rate of conversion of solar energy into wind energy is about 200 trillion watts (2×10^{15} W) for the entire world. This represents about 1000 times more power than is used by the entire world population. Perhaps the most notable feature of wind energy is that it is a renewable resource.

PRINCIPLES OF WIND ENERGY DEVICES

Winds can serve as an energy source because the air molecules in the atmosphere are moving in one general direction. A moving molecule has kinetic energy (K.E.) which can be described by the formula K.E. $= \frac{1}{2} MV^2$, where M represents the mass of the particles and V the velocity of the wind. The amount of air that encounters the blade of a windmill is equal to the product of the velocity of the air, the density of the air, and the area of the windmill blade. The power available from the windmill is the kinetic energy available divided by some unit of time, usually a second. Substituting these relationships in the kinetic energy formula, we obtain

$$\text{power} = \frac{\text{kinetic energy}}{\text{time}}$$

$$= \frac{1}{2} \times \text{density of air} \times \text{area of the blade} \times \text{velocity of wind}^3$$

The most important thing to note is that the power of a windmill is related directly to the cube of the wind velocity. This means that if the wind velocity doubles, theoretically the power available from the windmill will be increased eight times. Since the blade of a windmill rotates, the blade generates a circular surface that has a radius equal to the blade length. The area of a circle is equal to the product of π times the square of the radius (which in this case is the blade length), so if the blade length is doubled, the power available from a windmill could be increased by a factor of 4. Theoretically, only 59% of the energy from the air striking the blades of a windmill can be converted into usable energy. If all the energy from the wind were changed into usable energy, the air molecules would have no energy left to move away from the windmill. The actual efficiency of most windmills using 59% as the maximum value is between 50 and 80%.

USES OF WINDMILLS

Wind power has been used for centuries for moving ships over water (see Figure 13-2). Other early applications included using windmills to pump water and turn millstones to grind grain. The windmill probably originated in Persia during the seventh century. They have been used in Europe since the eleventh century, with the most notable concentration of windmills being in the Netherlands (see Figures 13-3 and 13-4). Here they were used extensively to pump water from land that was below sea level.

Windmills in America were used predominately to pump water on farms, as shown in Figure 13-5. They were also used as a source of electric power in rural areas in the 1920s. At that time there were more than 20 windmill manufacturers in the United States with investment capital of several million dollars. In 1930 the Rural Electrical Administration began to provide cheap electricity to farms from conventional sources, fossil fuels and hydroelectric generators. This was a cheap and more reliable source of electrical energy. The American windmill industry nearly disappeared.

Figure 13-2 Small fishing boat near the coast of Denmark.

Figure 13-3 Windmill in the Netherlands that was used to grind grain.

Figure 13-4 Windmill in Germany that is used to produce electricity.

With this decline in interest in small windmills in the 1930s, attention was shifted to the construction of a few large-scale wind generators. These range in scale from 90 kW to a 1.25-MW generator built at Grandpa's Knob, Vermont. This windmill has a 110-ft tower and two 90-ft blades. In 1945 it was part of the electrical system of that region,

Figure 13-5 Windmill in the Midwest that was used to pump water.

but one of the blades broke and the windmill was subsequently abandoned. Today some people envision fields of windmills on 800-ft
towers, each capable of delivering 4 MW of power dotting the landscape. About 250 of these, hopefully silent, giants would be equivalent
to one large nuclear power plant. One problem with large windmills
today that was not a problem in the 1930s is that they can interfere
with television reception.

Projections for energy sources in the year 2000 include estimates
that 5 to 10% of total U.S. energy needs will be coming from wind
power. Since wind velocities in the United States average 10 to 25 miles
per hour only about 2 days per week or less, most windmills will be
operating at 20% of their load capacity the majority of the time. This
means that wind power will at best serve as only one of several electrical
energy sources in a particular area.

Perhaps the most popular use of windmills will be made by re-
verting back to using them as a source of electrical energy for single-
family dwellings, particularly in rural areas. The workhorse of small-scale
wind devices has been the Jacobs windmill. Their capacity is 1500 W,
and they could be expected to generate 400 to 500 kW per month.
The Jacobs company went out of business in 1957, but not before
selling over 50,000 electric generators. Today a windmill with a 10-ft
blade could supply the electricity needs of many families in some parts
of the country (see Figure 13-6.)

By far the most economical use of a windmill occurs when the
electrical load requirements are large enough so that the maximum
electrical energy production the windmill can produce at any time can
be utilized by the system. This application does not require an energy
storage system. An example of this type of implementation is a modern

Figure 13-6 Windmill with a vertical shaft.

dairy farm, where electric motors to run milk coolers, water pumps, and other devices are nearly always in operation.

One of the fortunate aspects of wind power is that in many instances it can be used to complement solar heating. The wind often blows at night and on cloudy days. If electrical heating is used as a backup to solar heat, electricity produced by a wind generator would provide a complementary heat source from a renewable energy supply. When solar collectors falter during winter blizzards in cold climates, windmills will be operating at maximum efficiency.

ENERGY STORAGE SYSTEMS

Electrical energy produced at an inappropriate time can be stored in several ways. One method is to store the electrical energy directly in one or more batteries that can be discharged when energy demand is high or wind power is not available. However, batteries are expensive and energy is wasted because the charging and discharging processes are not very efficient.

Electrical energy from windmills can also be used to pump water into a container that is located several feet above the ground, such as a water tower. When electricity demands are high, this water, which now has potential energy due to its elevation, can be allowed to return to the surface elevation. On the return trip the following water can turn a water turbine which rotates a generator to produce electricity. This method is already being used by a conventional electric power utility generating plant on Lake Michigan, where lake water is pumped above the lake level into a pond behind a dam. This allows the power company to run their heat-generating equipment above the level required when system demand is low and below the maximum level required when system demand is high. In this way they are obtaining better use of a power-generating facility that may ordinarily run at peak capacity only for brief periods each day.

Another possible use of electricity generated by a windmill would be to break down water into hydrogen and oxygen. This reaction requires energy. The hydrogen is stored and used as a fuel. When hydrogen burns, it combines with the oxygen in the air to produce water. As in any other energy conversion system, some energy is always lost as waste heat, but the production of hydrogen has some advantages in that it can be transported and used anywhere as a nonpolluting fuel. It can also serve as an alternative to gasoline in automobiles. This would be one way to utilize energy produced by large windmills that could float on the ocean several miles from land, so that it would be difficult to transfer the electrical energy produced using conventional electric power lines.

QUESTIONS

1. Why is precipitation often associated with a low-pressure system and fair weather often associated with a high-pressure system?

2. Many windmills have a braking system or a blade feathering system that causes the windmill to stop functioning in very high winds. Why is this necessary?

3. What present technological development, which did not exist in the 1920s, is affected by large windmills?

4. Windmills provided the electrical needs of many farm families in the 1930s. Cheap electrical energy from other sources replaced most windmills in the 1940s. If electrical energy prices were to rise dramatically in the next few years, why can't we return to using windmills to meet all our electrical energy needs?

5. Why does the addition of an electrical energy storage system significantly reduce the efficiency of windmill?

6. How can solar and wind systems together partially compensate for deficiencies of each separate system?

7. If you want to know about the dependability of significant winds in your locality, what kind of hobbyist should you contact for advice?

8. Wind power that is stored as electrical energy in a battery furnishes direct rather than conventional alternating current. What kinds of devices in your home could function on either alternating or direct current?

Biomass: Let Nature Work for You

Biomass, plant material in any form, has significant potential for use as a supplemental fuel in the future. Biomass can be used directly, such as the combustion of wood or other crops grown for their fuel value, or it can be converted to other forms, such as ethanol or methanol for use in gasohol. The wastes from various human processes and municipal refuse can also be burned directly or converted to other fuels.

WOOD

Wood was the primary source of energy in the United States until the late nineteenth century. Since the 1973 Arab oil embargo, 7% of Americans have switched to burning wood in stoves or fireplaces, at least as a supplemental fuel. What is the likelihood of trees becoming a major energy source? To answer this question it is necessary to look in detail at the quantities of wood required and the available land area necessary to support such an industry.

Wood, because it is primarily cellulose, has an advantage over other proposed energy feedstocks in that there is no competing use for it as food. It can also be grown on land that is marginal for agriculture, such as erosion-prone hillsides and barren land under power lines, while enhancing the beauty of the surroundings. It should be noted, however, that growing wood on marginal lands will require costly fertilization, irrigation, and the possible development of new tree species. It is estimated that the United States used approximately 1.5 quads (10^{15} Btu) of wood energy in 1980 compared to a total U.S. energy consumption of 80 quads. Projections by the U.S. Office of Technology Assessment suggest that wood might be able to supply up to 10 quads by the turn of the century.

For these hopes to be realized, proper forest management is essential. Over the past decade the demand for firewood has led to massive deforestation in many parts of Asia and sub-Saharan Africa. Even in the United States, the actual acreage of commercial forestland is constantly being reduced (Figure 14-1). The gap between wood removal and growth is continuously decreasing. Another consideration is the type of wood. Not all tree species are equally desirable from a Btu standpoint. For example, aspen has only 59% of the heating value of an equal volume of white oak. The major determining factor is the moisture content. However, there are better uses for scarce hardwood than burning it. Rapidly growing softwoods such as poplar would be a better choice for

Figure 14-1 The actual acreage of commercial U.S. forestland is constantly being reduced.

a fuel. It takes a hard wood such as oak 100 years to reach the size that a poplar attains in 10 years.

About 10 acres of land is required to provide a continuous supply of wood to heat a home in the northern part of the country. Using Wisconsin as an example, this means that 12 million acres of commercial forest would be required to supply the space heating demands of its 1.2 million single-family dwellings. Although Wisconsin is known for its forests and its paper industry, this amounts to 82% of all forestland now existing in the entire state. Even if unproductive acreage were brought up to an average growth rate, it is unlikely that more than 0.02% of the state's total energy demand could be provided by wood.[1]

The burning of wood also poses environmental difficulties. Many New England cities develop a wood-smoke haze in winter. Vail, Colorado, has limited new homes to one wood stove apiece. Wood fires are totally banned in London.

The commercial/industrial use of wood and wood wastes is, however, increasing. The U.S. pulp and paper industry now satisfies about 50% of its own internal energy demand by the use of wood wastes. Burlington, Vermont, has built a 50-MW wood-chip furnace to generate electricity for its 20,000 residents. About 5% of industrial boilers are now designed to use wood.

OTHER ENERGY CROPS

A number of other crops are now being investigated which might also serve as energy sources for the future. Green and blue-green algae are being artificially (and expensively) farmed. Because of their high protein content, however, it may be more appropriate to use them as a food source. Water hyacinths, originally studied as a potential method of water pollution control, can proliferate in sewage and can serve as raw material for generating methane gas. Kelp, a marine perennial, is also under investigation as a raw material for methane production. It grows rapidly in the ocean and additionally, enhances the local fish population by serving as a food.

Sunflower oil is an attractive alternative to diesel fuel. Full-scale tests of this oil are being undertaken in South Africa and Australia. U.S. soybean and peanut farmers are investigating the possibility of converting portions of their crop to fuel production. Such use will, however, compete with the food supply.

More exotic crops, suitable for growth in arid, marginal areas, in-

[1] Ray Young, Dept. of Forestry, University of Wisconsin, Madison, private communication.

clude buffalo gourd (*Cucurbita foetidissima*) and gopher plant (*Euphorbia lathyris*). Both of these plants have the advantage of flourishing in areas that are unsuitable for other crop production. Buffalo gourd can produce up to 2 barrels of seed oil per acre (about 10^7 Btu per acre). This is more fuel per acre than the sunflower can produce. In addition, a root starch is produced which could be used for alcohol production. Approximately 3.5×10^7 Btu per acre per year could be generated from this material by conventional ethanol fermentation. The buffalo gourd is indigenous to Arizona, Texas, and New Mexico. Gopher plant, a biennial shrub indigenous to California, produces both an extract similar to gasoline (5% of dry weight) and sugars that can be fermented to alcohol (20% of dry weight). At a biomass yield of 8 to 10 dry tons per acre per year, this corresponds to about 3 tons of useful fuel per acre. It is estimated that up to 4.5×10^7 Btu per acre per year of liquid fuels can be produced from this material.

At the other extreme in growing conditions, the Chinese tallow tree (*Sapium sebiferum*) is a prolific oil and fat producer which thrives on coasts and in a number of Asian countries (India, China, Taiwan, Pakistan). The seeds are coated with a semisolid tallow; inside is an oil similar to linseed oil. The yields of each of these are typically 6 barrels per acre, providing a total fuel value of 6 to 7×10^7 Btu per acre. Because the tallow tree grows very rapidly, it may also serve as a fuelwood crop.

Although the oilseeds and related crops offer the potential of supplementing our fuel resources, much research remains and the economics, land-use issues, and possible fuel–food trade-offs are debated.

GASOHOL

Gasohol is a combination of gasoline and an alcohol (either ethanol or methanol). Either alcohol, which can be used in percentages of up to 10 to 15% without requiring engine modifications, is available from nonfossil fuel sources, and hence can be used to extend our conventional petroleum reserves.

Fuel-grade alcohol plants are not new. A number were designed and built in the United States during World War II. Several that use process wastes, that is, wastes generated in industrial processes, as the raw material are still operating in the United States and Canada.

Ethanol is fermented from corn or other grains, sugarcane (particularly in Brazil), or potentially from a variety of organic waste products. The sugar in these materials is what is converted to ethanol. Methanol can also be produced from waste products, wood, or coal.

One of the major difficulties with gasohol use is that if the alcohol is produced from a crop such as corn, the energy output/input ratios are not favorable. More energy is required for seeding, fertilizing, irrigating, and harvesting the corn than can ever be regained in gasohol consumption. Furthermore, competition for the use of grain for foods is another consideration. The energy economics, however, are vastly improved if waste products serve as the raw materials.

Much recent interest has been focused on the conversion of cellulosic wastes to ethanol. The fermentation of cellulose is more difficult than that of sugars, requiring some pretreatment (soaking, grinding, alkali impregnation) followed by hydrolysis to convert the cellulose to sugars prior to fermentation. Cellulose cannot be fermented directly; however, considerable cellulosic wastes, as well as timber, are available. The paper industry, for example, uses only 40% of the potentially available wood for pulp production. The wastes, in the form of sawdust and wood chips, could be used for ethanol production. In addition, the processing liquors, the solutions used to convert the raw wood to pulp suitable for paper manufacture, could be converted to useful chemicals. The potential for production of fuels from paper industry wastes is great. Recently, a spokesman from Exxon stated:[2] "The world will move to non-depleting, renewable resources such as biomass, with the paper industry leading the way."

A number of proposed methods for the commercial production of ethanol from biomass are under investigation. If the raw material is cellulosic, sodium carbonate or calcium hydroxide may be used to swell the materials, making them more susceptible to further treatment. Most raw materials are then treated by one of the following processes:

1. Conventional pulping processes utilizing higher temperatures and pressures than normal
2. Nonconventional pulping techniques such as those utilizing acetic acid or alcohols
3. High temperature and pressure
4. Concentrated acids such as sulfuric or hydrocholoric
5. Microbial digestion

All these will produce glucose, a sugar, which can then be converted to ethanol by the following fermentation reaction:

$$C_6H_{12}O_6 \xrightarrow{\text{enzymes}} 2CH_3CH_2OH + 2CO_2$$

(glucose)   (ethanol)   (carbon dioxide)

[2] "Exxon Forsees Oil Imports Peaking in 1990," *Chemical Week*, Vol. 129, No. 23, June 10, 1981. p. 23.

The use of methanol obtained from biomass is also gaining considerable favor. A large wood-based production facility planned for the Northeast is projected to provide up to 3% of New England's motor fuel demand. The peat reserves in North Carolina are also being studied as a potential raw material. Much work has already been done in Sweden regarding the conversion of peat to methanol. Nearly 70% of Sweden's net energy consumption is supplied by imported oil (the remainder being primarily nuclear). For reasons of national security, Sweden wishes to rely more heavily on its own supplies of wood and peat.

As with most alternative sources of energy, the current limitations to methanol use are primarily economic, not technical. The transportation costs alone of the product can be 25% of the purchase price.

Although neither ethanol nor methanol have as much fuel value as gasoline, they can boost the power obtainable from the gasoline. Methanol, for example, has a "blending octane" of 120 compared to from 87 to 92 for most gasolines. They also reportedly allow smoother burning, resulting in better miles per gallon ratings than those of gasoline alone.

Methanol and ethanol blends are not without technical difficulties. Both show some material incompatibility with, for example, plastics. At low temperatures, the water normally present in gasoline can cause separation of the alcohol and gasoline into two layers, thus limiting their use in the colder climates. Blending agents can be added, however, to minimize these separation problems.

Pure methanol or ethanol can also be used as automobile fuel. Methanol has for many years been the fuel of choice of race car drivers. In Brazil, where ethanol can be competitively produced from sugarcane, many automobiles are now being modified to use ethanol exclusively.

There are also other possibilities for gasohol-like fuels. Blends of ethyl acetate, turpentines, and other organic chemicals with gasoline also show some promise and are being investigated.

REFUSE-DERIVED FUEL

Refuse-derived fuel, commonly called RDF, is produced from the combustible portions of municipal refuse. This refuse consists of 50% (shredded) papers, about 20% food wastes, and 5% each wood and plastic (see Figures 14-2 and 14-3). About 80% of this municipal waste is combustible. Every day each U.S. citizen produces about 3.5 lb of refuse. Thus significant quantities of these wastes are available.

Many municipalities are now separating their trash. The RDF can be burned in boilers for the production of electric power; the glass can

Figure 14-2 Collection of garbage and trash to be used as RDF. (Courtesy of the Electric Power Research Institute.)

Figure 14-3 Typical material received by a RDF-combustion facility. (Courtesy of the Electric Power Research Institute.)

theoretically be recycled, perhaps as filler in bricks or to make foamed glass, and the metals can be recycled or disposed of in a landfill.

A major difficulty with RDF is its lack of consistent fuel value. Because of the variation in heating content of different wastes, it does not make a suitable fuel for any process where temperature is a critical factor.

Trace amounts of toxic impurities can also create problems. When plastics burn, they generate more smoke than wood or paper. Moreover, some plastics, such as polyvinyl chloride (PVC), release hydrochloric acid during combustion. Hydrochloric acid is very corrosive and could do significant damage to materials and to plant or animal life if the stack gas were not handled properly.

Another limiting factor in the use of RDF is the energy necessary to process the wastes (Figure 14-4); to separate them, concentrate the combustible components, and to put them into a form suitable for easy combustion. Most householders do not wish to make the effort to classify their garbage. There is no real economic incentive for them to do so. This separation, therefore, must be done at the combustion facility.

Figure 14-4 Processing of RDF. (Courtesy of the Electric Power Research Institute.)

It is not an easy task, and (compared to the energy available in the RDF) requires that a good deal of energy be expended.

Since the energy value per pound of RDF is quite low, on the average, compared to the fossil fuels, a larger quantity of RDF is required to provide a given amount of energy. This further restricts their use. For these reasons, its adoption as a fuel has not been as rapid as one might hope.

Recent studies have investigated the feasibility of burning both coal and RDF simultaneously in conventional power plant boilers. Pyrolysis (heating the wastes in the absence of oxygen to produce a low-Btu gas and tarlike oil), hydrogenation (adding hydrogen by chemical means), and biological conversion of the coal to methane or alcohol are also currently under study.

Because the raw material is waste, its cost is essentially zero. In fact, money is saved by not having to dispose of the materials by landfill or by other conventional means. This is generally the major incentive for burning RDF. However, in those cases where the items could be reused, the greatest energy savings of all would be realized. Considerable energy (and money) have been invested in creating a usable product. Only if no further use can be made of the item is it truly beneficial to burn it for energy.

QUESTIONS

1. What are the limitations on again using wood as a major source of energy in the United States?

2. What crops might serve as useful alternatives to conventional fuel?

3. How is ethanol for gasohol likely to be produced? What is the starting material? What is the fermentation reaction? If cellulose is used instead to serve as the raw material, what preliminary steps might be used prior to fermentation to ethanol? What advantages are there to using cellulosic raw materials rather than starting directly with those composed of sugars?

4. What is RDF? Why is everyone not burning it to produce electric power? What is the major advantage of using it as a fuel?

CHAPTER 15

Other Alternative Systems: Are These Feasible Alternatives?

TIDAL POWER

With the exception of solar energy, the energy generated by the gravitational interaction of the earth, the earth's moon, and the sun is the only instance of a continuous impact on the earth by other members of the solar system that has significance to people. This gravitation produces the tides in the ocean. The tides occur every 12 hours and 24 minutes and range from 3 to more than 30 feet depending on location. The long-term impact on the earth is that the rotation of the earth and therefore the length of the day, is decreasing by 0.001 second every century. Although this shortening of the day appears to be trivial, it does involve more than 3 billion watts of power.

The most obvious method for using the tides as a source of energy would be to use the moving water to turn the blades of a turbine. However, the concentration of tidal energy is very small, so that millions of generators would be required and this is not economical.

An alternative model is to consider the movement of water in the oceans as being similar to the movement of water in rivers. River water is collected behind dams and when the water is released it drops to a lower elevation, thus changing the potential energy to kinetic energy, which in turn may be used by a turbine to run an electric generator. If the coastal geography is appropriate, a similar facility can be made for collecting water at high tide and releasing it at low tide. This concen-

Figure 15-1 The Bay of Fundy in eastern Canada, where the tide changes by more than 40 ft.

trates the energy enough to make it economical. A tidal generating plant has been constructed in France which can generate 240 mega-watts of power. An examination of the most reasonable sites for using tidal power in this manner was not very promising. It appears that less than 5% of the electrical needs of the developed countries can be sup-plied using this technique. Several individual sites have been studied, including the Bay of Fundy in eastern Canada, where tides are more than 40 feet high (see Figure 15-1). In each case, however, generating electricity using the tides is still too expensive compared to conven-tional electrical energy sources.

OCEAN THERMAL ENERGY CONVERSION

The oceans cover more than 70% of the earth's surface and therefore they receive about 70% of the solar energy that strikes the surface of the earth. Water can absorb or release large quantities of heat with small changes in temperature. This means that the oceans and other large bodies of water have a stabilizing effect on the air temperatures over or near them. The surface water of the oceans is constantly in motion due to the winds above the surface and rotation of the earth. The ocean water below the surface is more dense and several degrees colder than the surface water. In some cases this temperature difference is nearly 50°F. A temperature difference between two regions indicates a difference in the concentration of thermal energy. A difference in energy concentration means, according to the second law of thermodynamics (Chapter 2), that energy will flow spontaneously from the region of higher energy concentration (the warmer region) to a region of lower energy concentration (the cooler region). A spontaneous energy flow can then serve as an energy source for activities other than heating cold water.

Energy from the oceans is often described using the term OTEC, which is an acronym for "ocean thermal energy conversion." The OTEC model for generating energy reads a little like the stories that consistently alternate between "good news" and "bad news."

The good news about OTEC contains the following elements. There is plenty of energy available; some estimates are as high as 20,000 MW by the year 2000. The modifications of overall ocean temperatures are probably less than 0.0006% on a worldwide scale. At a specific location in the Gulf of Florida, the Gulf Stream could provide 3 million megawatts of electrical power. This energy, at 2% efficiency, would be extracted from water moving at 1.5 billion cubic feet each second in a path 20 miles wide and 500 miles long. The temperature difference at the present time between the surface water and the under-sea water at a depth of 2000 feet is $45°F$. Removal of the energy described would reduce the temperature difference to about $43.5°F$, which does not appear to be a significant factor. This could produce 3 million megawatts of electricity or about 10 times that of all the electricity produced by electrical utilities in the United States in 1981.

The bad news relates to geography and cost. Energy generated by OTEC operations will all be done in the oceans, which means the energy must be transported to population centers. The Gulf of Florida is noted for high winds and rough seas, particularly during the hurricane season. Direct transmission of electricity from the generating station to land does not appear to be practical. One alternative is to convert the energy into a transportable form such as hydrogen, derived from decomposition of water. The hydrogen would be transported and used as a fuel on land.

A second possibility would be to use the electricity produced by the OTEC facility to produce (at sea) useful but highly energy intensive substances such as ammonia and aluminum. Ammonia, which is made from nitrogen and hydrogen, can be easily synthesized at sea because nitrogen is available from the air and hydrogen is available from the water. This is also bad news because there is significant transportation expense involved in bringing aluminum ore to the OTEC site. The good news is that materials can be transported over the ocean more cheaply than by any other means.

OTEC technology is based on using the same operating principles as those of a refrigeration unit. A fluid condenses to form a liquid in the low-temperature region, then boils to form a gas in the high-temperature region. The expanded gas volume is used to turn a turbine, the turbine turns a generator, and electricity is produced. Because of the small temperature difference between the warm and the cold regions, the overall efficiency of the system could not exceed 7%. Most calculations of energy output are done, assuming operating efficiencies of about 2%.

The cost of producing electric power using an OTEC system appears to be about twice the cost of coal-fired or nuclear plants. As with many other forms of alternative energy technology, if traditional fuel prices double, economic feasibility may become a reality. One advantage of OTEC systems is that they all inhabit similar environments, which means that they could be mass produced. This would significantly reduce production costs and production time.

GEOTHERMAL ENERGY

Geothermal energy has the distinction that it is not associated with solar energy. The energy source is the interior of the earth (see Figure 15-2). The inner core of the earth, a region approximately 1400 miles across, is probably a superheated solid, that is, a solid that is actually at a temperature above the normal melting point of the solid but due to the intense pressure of surrounding material cannot expand to become a liquid. The outer core of the earth, which is a concentric sphere surrounding the inner core, has a thickness of about 1500 miles and consists of molten liquid material. The mantle, which is located directly below the crust, is a solid material about 1800 miles thick. The earth's crust, which makes up the outer surface of the earth, varies in depth from 5 to 20 miles.

The heat from the molten material in the core is not directly accessible to people and the thermal energy transmitted to the surface from these great depths is so diffuse that it cannot serve as a useful energy source. There are, however, isolated pockets of molten material located from 2 to 3 miles below the earth's surface that are useful energy sources. These regions are heated by energy released during the nuclear decay of radioactive elements and the pressure produced by miles of solid material above.

Groundwater can transfer this heat to the surface. When groundwater comes into contact with solid rock that has been heated by molten material, it may be heated to a temperature above its normal boiling point. This means that if there are sufficient cracks or porous material above the water, it will expand and travel to the surface where

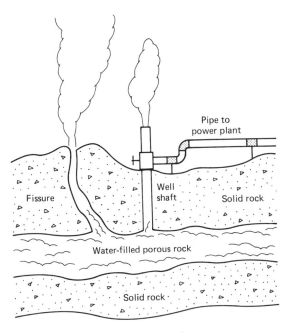

Figure 15-2 Diagram of the components of a geothermal energy site. (Courtesy of the U.S. Department of Energy's American Museum of Science and Energy, operated by Science Applications, Inc.)

the heat is quickly released into the atmosphere (see Figure 15-3). This release of energy escape is not useful for people who wish to use the geothermal energy. If, however, the heated water remains trapped below the ground, it can be channeled into an appropriate device, such as a turbine, and electrical energy can be produced (see Figure 15-4).

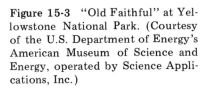

Figure 15-3 "Old Faithful" at Yellowstone National Park. (Courtesy of the U.S. Department of Energy's American Museum of Science and Energy, operated by Science Applications, Inc.)

Figure 15-4 Workers capping a geothermal energy source. (Courtesy of the U.S. Department of Energy's American Museum of Science and Energy, operated by Science Applications, Inc.)

Most of the known sources of geothermal energy are located near the Pacific Ocean. Other prominant locations include Iceland (Figures 15-5 and 15-6), Italy, and the southwestern portion of the USSR. At the present time two-thirds of the electricity produced using geothermal energy comes from generators in California and Italy.

The geothermal installation at The Geysers, California, located 75 miles northwest of San Francisco, presently produces about 800 MW of

Figure 15-5 Construction of a heat transfer line from a geothermal energy sources in Iceland.

Figure 15-6   Swimming pool in Iceland heated by geothermal energy.

power and will eventually be able to produce 2000 MW. At the present time the only cheaper source of electricity is hydroelectric power. A geothermal installation, however, does not require altering large amounts of land as is the case with hydro or solar energy sources, although The Geysers plant does require more than 3 square miles.

There are some problems associated with geothermal energy. One problem is that it cannot be made available everywhere. When all the obvious locations have been utilized, other, less desirable geothermal sources will be much more expensive. Also, this is not a pollution-free system. There are large amounts of wastewater that contain high levels of dissolved salts. In some cases this water may be returned to the earth to replace water that has been removed. This can solve a potential problem of land sinking and earthquake activity due to the removal of large amounts of water. The experience near Denver, Colorado, has been that injecting material into the ground has increased the frequency of earthquakes.

As has been the case with several other energy sources, the total amounts of energy available from geothermal sources are significant, perhaps even mind-boggling, but they are frequently not sources of concentrated energy that can be used directly in an efficient manner. The other problem is that geothermal sources of energy are not widely available.

QUESTIONS

1. Water turbines are used to produce electricity at hydroelectric power plants on rivers. Why aren't water turbines placed in shallow ocean water that is influenced by the tides not an effective means of generating electric power?

2. What kind of geographic feature on an ocean coastline would be particularly useful if a facility is to be built that generates electricity using tidal power?

3. How does a sustained difference in temperature in ocean water at different depths create a circumstance that allows for the generation of electrical energy?

4. What are several things you would want to consider in determining the desirability of a particular location as a site for an ocean thermal energy conversion plant?

5. What is one reason for the low efficiency of an OTEC power plant?

6. Is geothermal energy a pollution-free energy source?

7. What are some advantages and disadvantages of geothermal energy?

CHAPTER 16

Energy Conservation: The Best, Cheapest, and Quickest Source

Energy conservation has been called by some our best, cheapest, and quickest source of energy. It is, in fact, our past wasteful and inefficient use of energy that now provides us with tremendous opportunities to conserve energy.

To understand the magnitude of the amount of energy potentially wasted in the United States, let us consider two examples.[1]

1. About 20% of current white-collar workers in the United States are employed in managerial or administrative positions. Suppose that half of these people unnecessarily leave their office lights on for an hour during the day (perhaps during the noon hour). The energy this wasted would amount to 1000 megawatt-hours, which is equivalent to the output of a large power plant operating for an hour.

2. If one-half of U.S. households leave their TV sets operating for 15 minutes unattended, again the wasted energy is 1000 megawatt-hours.

[1] "The Second Law," Northwest College and University Association for Science, Richland, Wash., November 1977.

METHODS OF CONSERVATION

There are two general approaches to energy conservation: the technical fix method, and making changes in our life-styles. Both are necessary if we seriously wish to decrease our energy consumption.

The *technical fix* centers around developing more-energy-efficient technologies: better insulated homes, driving smaller cars, or converting to less-energy-intensive industrial manufacturing processes. Usually, these changes are a one-time event. That is, once they have been made, further changes to conserve additional energy are not likely to be successful, at least not until new technologies have been developed.

Efficiency, as we saw in Chapter 2, is a measure of the energy actually obtainable from any device (an automobile, appliance, or furnace) compared to the energy input. Although no energy can ever actually disappear, it can be "lost" for practical purposes by converting it to low-grade waste heat. The more efficient a device or process, the less energy is wasted as nonuseful heat. In the past, since energy was cheap (and getting cheaper), we paid little attention to the question of efficiency. Economics dictated that other considerations were of more importance. For example, if energy is cheap, why bother to insulate your attic thoroughly? Insulation was relatively expensive until recent fuel price increases. This attitude was also true in other facets of our society.

Another major factor in conservation is that of time and convenience. This is particularly relevant when one considers *life-style modifications*. It is frequently more difficult to convince people to adopt changes in their life-styles than to use more efficient technologies. Life-style changes frequently require what are considered to be major sacrifices. Carpooling, lowering one's thermostat, living in a smaller home, or simply buying fewer "luxuries" may be unattractive alternatives. Yet it is likely that all of these will be necessities in the near future for the majority of Americans.

Yet, if we compare our life-style to that of many Europeans, as in England, Denmark, West Germany, or Sweden, the life-styles are not that much different, although we use approximately twice as much energy per capita. Surely there are ways that we can conserve without tremendous sacrifices—without going back to "living in caves."

Overall, we have had some success in energy conservation. According to a 1981 study at the Livermore National Laboratory, which included adjustments for economic growth and other factors, the United States was then conserving energy at a rate equivalent to 4 to 5 million barrels of oil per day, or 10% of the energy used in the peak year of 1973.[2] More savings, of course, were possible and have since been made.

[2] Ibid.

Figure 16-1 illustrates the resulting revised demand projection esti- mates for oil consumption, according to Standard Oil Company (Indi- ana).[3] As can be seen, by 1980 the free world oil consumption was expected to decrease significantly.

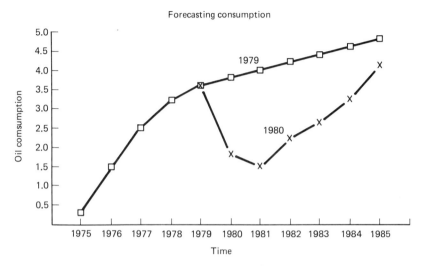

Figure 16-1   U.S. demand projection estimates for oil, 1979 and 1980 forecasts. [Courtesy of Standard Oil Co. (Indiana).]

RESIDENTIAL ENERGY CONSERVATION

How can energy be conserved around our homes? There are a variety of possibilities. To determine which approaches are likely to be the most effective, it is necessary to consider home heating, since it is the most energy intensive of residential demands. Since residential space heating requires such a large share of the personal energy budget, it is likely that the greatest savings can also be made in this area.

Some possible changes are fairly simple and inexpensive. For every degree the average temperature in the home is reduced, the fuel savings (in the northern part of our country) amounts to about 3%. Storm or plastic windows can cut heat loss through window glass by 50%. It is estimated that if only 1000 homes were better caulked, insulated, and weather stripped, 539 homes could be heated with the energy that is saved.[4]

Both technical fix and behavioral modifications can be used to re- duce the heat loss from buildings. The heat loss from a building is

[3] "Oil Supply and Demand," *Span*, No. 4, 1980, p. 16.

[4] Ibid.

proportional to both the temperature difference between the interior and exterior, and the total outside surface area of walls, ceiling, and floor. It is also a function of the construction materials, the insulation R values, and the window area. Heat losses can therefore be reduced by making homes smaller and by living in duplexes or apartment buildings, which share common (interior) walls. Reducing the window area reduces heat losses because of increased conduction of heat through glass areas compared to insulated walls. This is offset somewhat, however, by increased lighting requirements and reduced solar heat input.

The *payback period* is the time necessary to recover the initial investment by the energy savings attained. Adding insulation to one's ceiling, installing storm windows and doors, and adjusting one's furnace for greatest efficiency are modifications which are economical to make; insulating the walls or replacing the hot water heater are too expensive compared to the potential savings. Most suggested energy-saving methods are fairly simple to effect. It must be noted, however, that retrofitting an older home is much more difficult and expensive than including the same energy-saving techniques in the original construction. Specific ways, and the amount of energy that can be conserved around one's home, are listed in Appendix E.

Behavioral fix modifications can also reduce the heat loss. The difference between the interior and exterior temperatures can be reduced by either moving to a warmer climate or by lowering the inside temperature. Generally, the latter is more feasible.

The method for reporting this temperature difference is calculated by using *degree-days*. Heating degree-days are calculated by subtracting the average outside temperature (in degrees Fahreinheit) from 65°F (the temperature above which interior heating is assumed not required) for each day of the heating season and then multiplying by the length of the heating season in days:

$$\text{degree-days} = (65°F - T_{\text{ave out}}) \times \text{length of heating season}$$

Thus the total typical degree-days for Minneapolis, Minnesota, are 7853, but only 2826 for Atlanta, Georgia. The information as to the number of degree-days, plus the average length of the heating season, can be used to calculate the energy savings potential by lowering one's thermostate. Minneapolis, for example, has a heating season of 261 days. The average outside temperature can be calculated from the equation above, or degree-days = 7853 = $(65°F - T_{\text{ave out}}) \times 261$. The average outside temperature, therefore, equals 35°F. The original temperature difference (70°F − 35°F) is 35°; after lowering the thermostat to 65°F, the difference is only 30° (65°F − 35°F), a decrease of 5°. Since the fuel requirements are directly proportional to the difference between the interior and exterior temperatures, the energy savings is 5°F/35°F =

14%. A similar calculation for Atlanta would show that lowering the thermostat to 65°F would result in a 25% energy savings.

Other life-style changes may be more difficult to implement. Brief consideration of our European counterparts shows some differences, which we may have to try to copy in the future.

Many European homes (Figure 16-2) are built contiguous to one another, sharing common walls, similar to older "row houses" found in some areas of our country, such as in Baltimore. These homes are also tall and narrow. Frequently, European homes have less floor space than those in the United States, and many are multifamily dwellings, particularly in larger cities. These two factors can result in large fuel savings but are difficult to implement in the short run in the United States since housing patterns do not change rapidly.

Figures 16-3 and 16-4 illustrate other energy-saving features. The majority of the homes in Denmark are equipped with skylights which provide solar energy and light in the colder months, and can be opened for ventilation during warmer weather. Many Swiss homes, on the other hand, are built with adjoining barns, reducing heat loss through the walls.

Americans have had some success in household energy conservation. In the northeastern United States, for example, energy savings on oil consumption for central heating in 1980 ranged from 15 to 25% of that in 1973. Growth of electrical energy consumption by home appliances has also slowed dramatically during the past decade.

Figure 16-2 European homes typically are tall, narrow, and built immediately adjacent to one another (Brugge, Belgium).

Figure 16-3 Many homes in Denmark are built with skylights to provide both solar energy and ventilation.

Figure 16-4 Swiss homes are frequently built adjacent to barns.

TRANSPORTATION ENERGY CONSERVATION

Twenty-five percent of U.S. energy demand is used to provide transportation. Smaller and lighter cars with better miles per gallon (mpg) ratings are one alternative. Many people are now seeing the value of such cars. Other life-style changes, however, can also greatly decrease gasoline consumption. A list of these is given in Appendix D.

It is in transportation, however, that we can see the greatest contrast with Europeans. First, we simply have more automobiles and drive them more. Our driving habits are due partly to the large geographical size and distribution of population in our country, but another major contributing factor has been our dependence on cheap gasoline. In almost all other countries, there is a greater dependence on bicycles, taxis, and other forms of mass transit for passenger travel and for hauling of agricultural products and other freight (see Figures 16-5 to 16-14).

Several European cities have essentially eliminated personal cars. In Rotterdam, the Netherlands, 48% of all local trips are by bicycle. Runcorn, England, a town of 70,000, and Port Grimand, France, were both planned to function without autos. All transport is by bus, bicycle, or pedestrian walking except for emergency and special-delivery vehicles.

One of the common "solutions" that can lead to a decrease in energy transportation consumption is to improve mass transit in the United States. Buses and trains are significantly more energy efficient than automobiles in terms of energy use per passenger. However, the solution is not as simple as creating mass transportation. Our settlement patterns and past governmental policies of taxation, regulation, and subsidy have made it very difficult in this country to shift rapidly to mass transit. In the United States, automobiles are invariably required

Figure 16-5 Bicycles are the main method of transport in many European countries (here, Groningen, the Netherlands).

Figure 16-6 Major bicycle paths, such as this one in Denmark, cross many countries.

Figure 16-7 Bicycles are also used to haul produce (here, in Switzerland).

for transit to and from the railway station. In addition, the automobile is now such an integral part of U.S. life-styles that people will probably endure significant economic sacrifices to retain the use of their private vehicles. Taxis are not considered convenient except in very large U.S. cities. For longer trips, the airplane's increased speed will probably

Figure 16-8 Those who cannot ride bicycles for some reason find other ways to get around, as illustrated by this gentleman in Kristinehamn, Sweden.

Figure 16-9 Horse-drawn vehicles are still prevalent in Ireland for hauling milk and similar products.

Figure 16-10 Many European streets are exceedingly narrow, as illustrated by this one in Stockholm, requiring that cars be small and narrow.

Figure 16-11 Mass-transit vehicles must also be narrow to be accommodated by the narrow streets, as seen in this photo of an Amsterdam trolley.

Figure 16-12 Taxis are a very prevalent method of transportation in London. Private cars are not common and often not practical.

continue to outweigh the advantages of other means of travel in our fast-paced society.

In most parts of the world, automobiles are much smaller and more energy efficient than U.S. automobiles. The streets are frequently significantly narrower, requiring smaller cars, buses, and streetcars. For many years gasoline has also been expensive in places such as Western Europe and Japan since it is almost entirely imported.

Figure 16-13   Rapid, efficient rail service is prevalent throughout much of Europe.

Figure 16-14   Freight can be transported quite efficiently by barge. This is very common in the Netherlands.

The attitudes of the United States and Japan toward energy conservation differed dramatically until very recently. When the environmental movement began in the late 1960s, the first reaction of U.S. car manufacturers and the U.S. government was to install pollution control equipment on existing automobile designs. This equipment further reduced the mileage ratings of our automobiles. In Japan, however, the solution to reducing pollution was to use a more energy efficient automobile. Thus, not only were their new automobiles producing less pollution, but simultaneously, they were getting better mileage. In the United States increased mileage was a secondary consideration and tackled only piecemeal, and not seriously until the late 1970s.

As shown in Chapter 5, the overall efficiency of a typical automobile is very low. The internal combustion engine is capable, at best, of approximately 25% efficiency. For less-than-optimum conditions, the efficiency is considerably less.

In an effort to encourage gasoline savings, the federal government instituted a "gas guzzler" tax on 1980 model cars. In 1980, manufacturers whose cars were getting less than 15 mpg were required to pay taxes ranging from $200 (for those getting 14 to 15 mpg) to $550 (for less than 13 mpg). By 1986, the tax penalty ranged from $500 (for 21.5 to 22.5 mpg) to $3850 per automobile (for less than 12.5 mph).

A number of new automobiles, using new fuels, may be available in the future. Some possible new sources of energy are synthetic fuels, hydrogen, and electricity.

Synthetic fuels are those liquid or gaseous fuels made from substances such as coal or biomass. An example is methanol (or methyl alcohol), which is compatible with all automobile engines and burns cleanly. It does have a lower energy density than gasoline, so fewer Btus are available per pound of fuel; hence the tank has to be $2\frac{1}{2}$ times larger to give the same driving range. It can also be blended with gasoline, in percentages up to 15%, to produce gasohol. Ethanol (ethyl alcohol) can also be used to produce gasohol.

Hydrogen is also an attractive fuel for the long term. It has more energy per ounce than any other fuel known, it is abundant in nature, it could be a by-product of nuclear reactors, it is compatible with all types of automobile engines, it has very few emissions, and its sources are potentially inexhaustible. It is estimated that a current model automobile engine could be adapted to operate on hydrogen and oxygen for less than $1000. The source of this hydrogen and oxygen could be water. Water will decompose into these gases when an electric current is run through it. This could also be done within the engine. Liquid hydrogen must be stored at low temperatures, and this does not lend itself to use in the automobile. Hydrogen can also be absorbed on the surface of a metal for safe storage and released by a trickle charge of electricity, thus forming another type of battery within the car.

Electric cars are another possibility for the future, particularly for local trips. Electric cars are those vehicles which are powered by an electric motor drawing energy from rechargeable storage batteries or other portable sources. Today's electric cars can be driven 25 to 50 miles before recharging, at speeds up to approximately 50 mph. However, with present lead batteries you have to buy a new set every few months at a cost of about $800.

A closely related possibility is a hybrid vehicle, powered by a combination of a battery-powered electric motor and an internal combustion engine. The internal combustion engine adds range and acceleration, allowing the electric components to be used when they are the most beneficial.

Electric vehicles available today rely almost totally on lead-acid batteries similar to those now used to start cars. Their energy capacity and weight limit the range and increase the cost of the electric vehicles. Use of an electric car also increases the demand for electricity needed to recharge the batteries. This is attractive when considering our supplies of oil, but less so when one realizes that the production of electric power from fossil fuels is only 33 to 40% efficient.

COMMERCIAL ENERGY CONSERVATION

Because commercial energy demands are quite similar to household energy demands, the potential for conservation is also similar. Many stores, restaurants, and the like could lower their thermostats during the winter and raise them during the summer with no detrimental repercussions. People can learn to dress properly for the season of the year.

Since the mid-1970s, businesses have made some major changes in order to conserve energy. Extensive lighting for advertising is generally a sight from the past. The technology for controlled lighting systems is available for both retrofitting and new construction. Fluorescent lighting can be dimmed to less than 30% of the maximum light output, with a corresponding electrical energy savings. This allows different areas (hallways, conference rooms, office areas, school classrooms) to have different light-level settings. New stores are frequently constructed with less window area. This change not only reflects a need to reduce heat loss, but also changing U.S. life-styles. Seldom are potential customers lured into a store by an appealing display. Store hours may be shortened, at least in the short term, in a further effort to reduce energy costs.

INDUSTRIAL ENERGY CONSERVATION

There is a significant potential for decreasing industrial energy consumption by technical fix methods. Unfortunately, many of these methods require major capital investments which are impractical, sometimes economically impossible, to undertake for energy conservation reasons alone. Even in the most energy intensive industries, energy costs seldom comprise more than 2 to 8% of operating costs and only a few industrial sectors (chemicals, petroleum, and transportation equipment) are in a position to invest significant capital in energy conservation. Process modifications are feasible when a major change is required for other reasons or in the construction of new plants. This means that significant changes can occur only over a long period of time.

Figure 16-15 demonstrates another limiting factor due to the fairly direct correlation between a country's gross national product (GNP) and its total industrial energy consumption. The GNP is the total value of a country's annual production of goods. Although there is not a linear correlation, generally an increase in GNP accompanies an increase in energy consumption.

Economics and governmental policies have continued to hinder the potential energy conservation that might be attained by recycling industrial materials. Aluminum, for example, can be produced from scrap at approximately 5% of the energy required to produce it from ores; nonetheless, government subsidies such as depletion allowances have maintained the economic advantage of using these virgin materials. The same is true, although not so dramatically, for almost all manufactured goods.

One change that has occurred is industrial *cogeneration*, where the excess heat produced by one process (for example, electric power generation) is used for another purpose (such as heating buildings). This

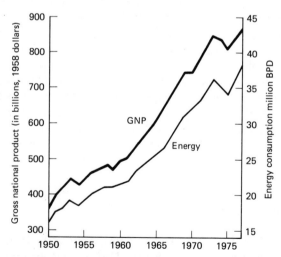

Figure 16-15 Relation between gross national product (GNP) and industrial energy consumption. (Courtesy of the U.S. Department of Energy's American Museum of Science and Energy, operated by Science Applications, Inc.)

process continues to have great potential as energy prices rise, for many industrial processes generate tremendous quantities of waste heat. This additional energy can be used on site or it can be sold to other companies. As an additional benefit, cogeneration will reduce thermal pollution of streams.

During the decade from 1972 to 1981, industrial energy consumption has averaged a 2.25% decrease while the output in manufacturing rose more than 17%. The chemicals industry led this decrease, with a 3.37% increase in energy consumption while production output increased 28%. Overall savings for these industries corresponded to an equivalent 1 million barrels of oil per day.

During the decade of the 1970s, industry also developed a greater reliance on electricity. Natural gas and oil consumption decreased 14% and 5%, respectively. There were, however, no emerging trends toward direct substitution of coal for these other fuels.

The Department of Energy has predicted that industry could, however, be consuming 50% of U.S. energy by 2000 unless more process changes are made and the best available energy conservation technologies are utilized. The latter could, overall, increase energy efficiency by 20 to 30%.[5]

As energy and other resources dwindle, some predict that industrial energy conservation will occur through changes in our life-styles. They say that we will switch from a manufacturing economy to a more-service-oriented economy requiring more education, health care, and social services.

ENERGY CONSERVATION IN AGRICULTURE

American farmers also have energy-related problems. They are under enormous pressure from several directions. High productivity per acre has made it possible to produce more food on less land, but low prices for farm products have made it necessary for farmers to farm even more land in order to maintain a given profit margin. Using more land means using land of lower quality because the best land is either already being used or is no longer available because of urban sprawl or road building. Upgrading marginal land requires more energy because the production, transportation, and application of fertilizers are all energy-intensive processes.

Farmers have conserved energy primarily by increasing energy efficiency (see Figures 16-16 and 16-17). This has been accomplished in

[5] "Industry Continues to Cut Energy Demand," *Chemical and Engineering News*, January 12, 1981, p. 7.

Figure 16-16 Much of the farm labor in Europe is still done by hand, as shown by this German woman and hay drying near Oslo, Norway.

several ways. One method has been to replace gasoline-powered equipment with more efficient diesel fuel engines. Another method of conserving energy on the farm is to engage in "single-pass" agriculture.

Figure 16-17 Small-scale farm equipment is more economical and energy efficient to operate. This gentleman farms in northern Norway.

Normally, a field is prepared for planting by plowing the old crop residue under in the fall and then breaking the soil apart using a disk and leveling the field for planting using a drag. The actual planting operation may be the fourth or fifth time a tractor has been used on the field. As the name implies, the single-pass process covers over old crop residues, breaks apart the soil, and plants and fertilizes the new seeds in one step. This dramatically reduces fuel consumption in addition to reducing labor costs and soil erosion potential.

Alternative energy sources for farms have applications that are similar to their use in other sectors of society: passively heated solar buildings for young animals, windmills to supply electricity to run milk coolers, and using manure and waste biomass as a source of methane gas for fueling tractors and other equipment. Alcohol fuel produced on a farm is also cost-effective because neither the plant sources nor the food by-products have to be transported.

OTHER RESOURCES

Our dependence on other imported resources, in addition to energy supplies, is illustrated in Figure 16-18. For many of these vital resources, our dependence on other countries is even greater than for

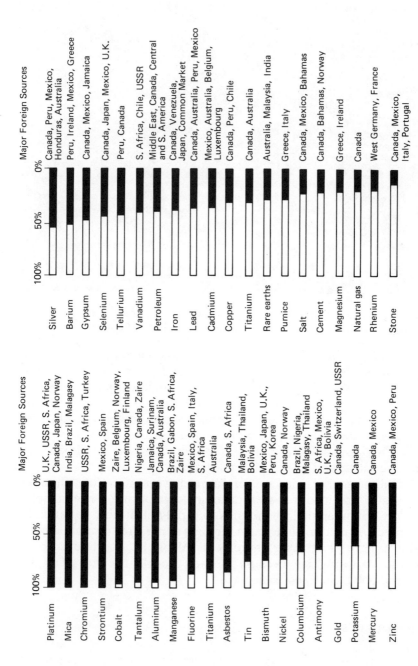

Figure 16-18 Percentage of U.S. mineral requirements imported during 1972. (Redrafted from "Curriculum Modules in Science, Technology, and Society," The Pennsylvania State University, with permission.)

petroleum. This is another concern, perhaps not yet fully realized, that we must confront during the next decade. Even if we could ever become energy independent, we could never become totally independent of other countries.

ADVANTAGES OF ENERGY CONSERVATION

To most people, the primary advantage of energy conservation is undoubtedly the personal monetary savings in their own pockets. This is particularly true as energy prices increase.

A decrease in energy consumption is, however, also a necessity from a national economic perspective. Even if adequate imported oil supplies were available through the late 1990s, it is unlikely that the United States could afford to pay for them. The balance of payments, the comparison of imports to exports, would have fiscal and economic impacts that would be unacceptable at all levels. A negative trade balance decreases the purchasing power of the dollar abroad. Eventually, the purchasing power would be so low that the United States could pay for nothing from another country. We are no longer the wealthiest country per capita as we were in 1970. In a decade, we dropped to ninth place. Even if these economic impacts could be minimized, in the future the internal and external political ramifications of the dependency would not be acceptable.

It is predicted that there are really two choices: to resolve the energy problem, particularly the petroleum import problem, gradually, by conservation now, or to settle it uncontrollably, through severe economic shocks to national economies, at a later date. However, in spite of the necessity of energy conservation from national and futuristic perspectives, the average consumer is likely to conserve energy only when economic savings can be made simultaneously. Human nature being what it is, very few people behave altruistically. Industries are similar: Their primary purpose is to make a profit. Investments that are not financially profitable generally will not even be considered. When establishing national policy, predicting energy demand, or estimating the market for a particular energy-saving device, this must be taken into account. Financial incentives must, of necessity, occur simultaneously with any other incentives. Only in this way will significant energy savings ever become reality.

Looking at the situation from another perspective, a little more altruistically, if we do not conserve now, particularly our fossil fuels, future generations may suffer not only severe shortages of fuel but also of raw materials for fertilizers. This would lead to shortages in food, pharmaceuticals, synthetic fibers, and other products which are now accepted as necessities in our daily lives. Then it will be too late to worry—the appropriate time is now, when supplies and alternatives remain.

QUESTIONS

1. Distinguish between the technical fix and behavioral fix methods of energy conservation. Which is easier to implement? Why?

2. Consult the local library, or contact your local weather station, to learn the typical degree-days and length of heating season for your community. How much energy can be saved by lowering the thermostat from $70°F$ to $65°F$? How much additional energy can be saved by lowering it further, to $60°F$?

3. Contrast life-styles in the United States and Europe. What modifications do you think you would be willing to make in your life-style in order to conserve energy?

4. Explain the likelihood of mass transit becoming an influential way of conserving energy in the United States in the near future.

5. What is cogeneration? How will it conserve energy?

6. For what other resources, besides energy, are we very dependent on other countries?

7. List a number of advantages of energy conservation. Which of these do you think are important reasons? Which of these would influence you either to make changes in your life-style or to make a financial investment for energy conservation purposes?

CHAPTER 17

Energy Economics and Policy: How Do We Get There from Here?

Economics is a major factor in a consideration of the U.S. energy situation. Low prices led to rapid growth of energy use during the 1960s and early 1970s; high prices have contributed to recent declines in the use of energy. The "supply" of energy resources is dependent on the price. Furthermore, the pricing mechanism is thought by many economists to be the best method for allocating our energy resources.

THE PRICE OF OIL AND THE ECONOMY

Figure 17-1 illustrates the rising price of crude oil and the corresponding increases in gasoline prices between 1973 and 1979. These increasing prices have played a major role in the high inflation in the United States during this period (Figure 17-2) and have contributed to the associated severe economic recession. Although the price increase in oil had been the most dramatic, other fuel costs had also risen during this period, adding to the inflation problem.

Conversely, the overall economy of a nation can affect prices. Figure 17-3 illustrates the effect of one type of major economic disruption, wars, on the price charged for crude oil. Large-scale economic and social upheavals such as these invariably influence supplies, demand, and consequently, prices. Relatively stable social conditions are generally necessary for prices to decrease, as seen in early 1983.

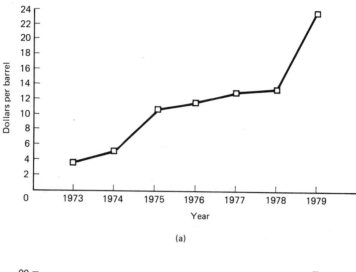

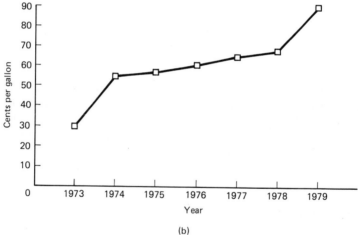

Figure 17-1 Rising prices of crude oil and gasoline, 1973–1979. (a) Dollars per barrel, benchmark Saudi crude oil; (b) gasoline price, cents per gallon. (Courtesy of the Society for the Advancement of Fission Energy.)

ECONOMICS AND ENERGY CONSUMPTION

U.S. energy consumption increased at a rate of 3.5% per year from the mid-1940s to the mid-1960s. As can be calculated from the doubling-time formula in Chapter 4, this corresponds to a doubling of our total energy consumption every 20 years. From the mid-1960s until 1972, the rate was 4.5%, which equals a doubling time of 15 years. Since our population did not grow nearly that rapidly, we must be using more

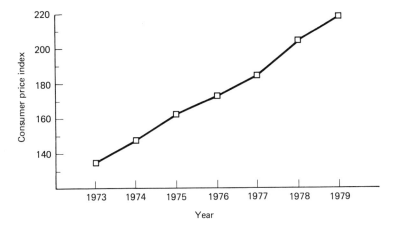

Figure 17-2   Rising U.S. inflation rate, Consumer Price Index. (Courtesy of the Society for the Advancement of Fission Energy.)

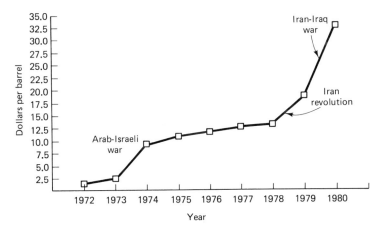

Figure 17-3   Effect of social conditions on crude oil prices. [Courtesy of Standard Oil Company (Indiana).]

energy per capita. During this period energy prices not only decreased but they decreased at an increasing rate. For example, during the period 1951–1971, in terms of constant dollars, coal prices decreased 15%, oil 17%, and electricity 43%. Plummeting electricity costs led to a sky-rocketing increase in air conditioning. Figure 17-4 illustrates the price changes during this period. As can be seen, prices decreased rapidly until the 1973 oil embargo.

It is easier to increase energy consumption when prices decline than to decrease energy consumption with corresponding price increases. Consumers frequently invest in very expensive appliances, cars, and

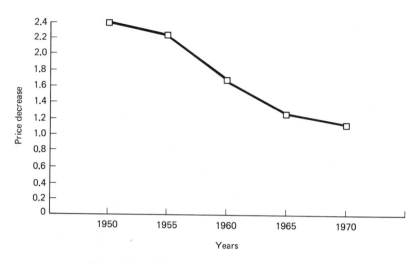

Figure 17-4   Energy price changes, 1950–1970.

other energy burners that have a long life and which are replaced at great cost. Once the item is purchased, consumers are locked into a fixed consumption level. Frequently, it requires much more energy to produce some items originally, such as power saws, electric knives, and garage door openers, than to operate them over their average life. From the consumer point of view, the fuel is only a small fraction of the total cost of owning and operating many items. Repair costs, finance charges, and depreciation usually are greater. The same is true from an industrial point of view. It is not economically feasible to purchase new equipment purely for the sake of energy conservation; moreover, the net result is likely to be an increase in overall energy use.

One current trend to reduce energy costs for both the consumer and the power companies themselves is *time-of-day pricing*. Electricity use generally peaks at specific times of the day (8 A.M. to 12 noon and 8 to 9 P.M.). It is to meet this peak demand that utilities must invest in more and more costly generating capacity. The costs of construction of new electrical generation facilities of all types have been escalating rapidly for a number of years. As discussed in Chapter 10, peaking facilities, which are designed to be used only intermittently, are also more costly to operate than base-load plants. If the demand for electric power can be maintained relatively constant over the course of a day, the overall electrical demand can increase, yet additional new generation capacity will not be required and the use of peaking facilities can be minimized. To decrease the daily fluctuation in electric energy demand, many electric power companies now charge, particularly their higher-demand customers, more for energy used during these peak periods than at other times. Individual consumers can also help in holding down their utility bills by using energy in off-peak times. How many times do our dishwashers operate right after dinner, when they could be used just as well at bedtime? When do we really need hot water? Couldn't a load of wash be done as well during the late evening news as at 8:30 A.M.? Although such changes would not totally eliminate the need for increased capacity (our population continues to increase), they could reduce the rate of expansion and hence the rate of increase of our utility bills.

ECONOMICS OF ENERGY SUPPLY

The amounts of energy resources available are dependent on the selling price of those energy supplies. Total energy reserves are, of course, fixed. However, the percentages of the reserves that can be extracted for use may vary with extraction cost. When prices are low, it is not economical to extract much of the proven reserves of oil, coal, natural gas, and uranium, but it is worthwhile when prices are higher. The higher prices of today have allowed energy producers to return to some of their previously abandoned sources, including coal refuse banks and oil fields, and to extract the additional quantities of these fuels which are now much more valuable, and therefore economical to recover.

There is a limit, however, to the percentage of the supplies that can be extracted economically. At some recovery level, the energy to recover the resource will exceed the energy value of the resource. When that limit is reached, there is no net energy gain. Further, recovery would not be feasible unless new technologies are developed using less-energy-intensive methods.

THE PRICING MECHANISMS OF CRUDE OIL

The pricing considerations for crude oil now include more than simple market forces. Direct involvement by governments in the international oil industry has brought new influences to bear on the market operations. These governments must also consider a wide range of economic, social, and political factors when determining policies. When establishing prices, therefore, national (government-owned) oil companies of the producing states have to take into account both governmental policies and commercial interests.

Crude oil plays a crucial role in the economies of most oil-producing countries. For the OPEC countries, for example, oil is their one major national resource. In most cases, the income derived from crude oil sales represents more than 90% of their total national revenues. The sale of petroleum is the selling of a national asset. From their perspective, it must therefore either be exchanged for another resource, or the income must be used to develop diversified economies that are healthy even after the oil reserves are depleted. Because of the eventual depletion of the oil reserves, oil-producing countries must balance current production rates with current and future economic needs of their country.

The majority of the world's crude oil never reaches the open market. Most of the trade is handled within the integrated operations of the major oil companies, in direct arrangements between producers and consuming governments, and with other users, such as refining and utility companies.

At one time the major oil companies controlled most of the world's oil supply. They were, to a large extent, able to control prices. However, during the 1970s the governments of the OPEC member states began taking control of production in their own countries. In 1970 the OPEC governments controlled only about 6% of their own production, but by 1981 ownership had risen to 88%. The major oil companies now own or have an interest in about 25% of the world's oil production.

As the influence of the major oil companies dwindled, the third-party buyers began purchasing directly from the producer governments or on the *spot market* rather than from the oil companies. Governments in some of the consuming countries, to assure their country of adequate oil supplies, have increasingly entered into direct government-to-government contracts with the producing countries.

The spot market, particularly the Rotterdam spot market, generally handles sales of crude oil in tanker-load or smaller quantities. It may also deal in oil products, particularly fuel oil and domestic heating oil. The supplies of crude for the spot market are generally those which are surplus to the requirements of a direct purchaser such as a major company. Companies short of crude will also deal on the market to make up their deficit. Trade on the market is handled by specialized traders who never physically handle the crude they are dealing in.

There is no one physical spot market. Traders can be located anywhere as long as they have good lines of communication. Since the bulk of the world's crude is shipped to the main refining centers, those destinations are used as reference points for quoting spot prices. For example, the majority of Western Europe's crude imports pass through Rotterdam; hence a Rotterdam spot market exists. Prices are quoted based on delivery from Rotterdam. Other major spot markets exist in Singapore, the Caribbean, and on the U.S. east coast.

There are many opportunities for speculators on the spot market. A given consignment has been known to change hands as many as eight times between being loaded for shipping at the producing country and arriving at Rotterdam, for example.

The spot market is the closest thing to an open market that exists for crude oil. Both consumers and producers use these prices as an important consideration in establishing their pricing. However, the prices on the spot market can fluctuate much more than other pricing. During periods of surplus, spot prices fall below the official prices set by the producers. During periods of shortage, spot prices can rise well above official prices, as companies that find themselves temporarily short will pay very high prices in order to keep their operations going. However, a long-term observation of spot market trends will reflect market development and can be used to predict future producer prices.

The volume of crude that is traded on the spot market can vary considerably. Many estimates put it as high as 10% of the total world crude sales, although during shortages it can drop to as low as 2% of the total.

ECONOMICS, PROFITS, AND THE OIL INDUSTRY

In recent years, the oil industry has been the recipient of many charges, including being monopolistic and deliberately causing the energy crisis in order to increase their profits. Are these accusations valid?

The oil industry is classed by economists as somewhat oligopolistic, not monopolistic. It is dominated by a few very large firms. The degree to which an industry is oligopolistic is determined by the percentage of the sales made by the four largest companies within the industry. In the oil industry, after the 1973 oil embargo, the four largest companies extracted a total of 31% of the crude petroleum and produced 33% of total U.S. refinery output. This is not exceedingly high compared to other U.S. industries, where 39% is typically controlled by the four largest companies within each industry. Furthermore, in 1972 there were 31 U.S. refineries with a production capacity of at least 50,000 barrels per day, compared to only 20 in 1951. The oil industry is, therefore, not heavily controlled by only one or two very large companies, and the industry has permitted growth by the addition of new companies.

It is true that some of the "energy crises" we experienced in past years could have been avoided. However, independent investigations have primarily blamed governmental regulations, particularly allocations, for these difficulties, not the industry.

Data obtained from a 1976 study by the *New York Times* illustrates that although oil company profits soared after the 1973 oil embargo, these profits started at a point well below the average for other U.S. companies, and ended only slightly above that average. Also, most of these high profits were from temporary factors, such as profits from inventory valuation and exchange rates.

Since that time, oil company prices and profits have increased further. This had been recognized by our government, which in turn passed the controversial 1980 Crude Oil Windfall Profits Tax.

ECONOMIC ANALYSIS OF ENERGY INVESTMENTS

There are many technological advances that could be implemented either to conserve energy or enhance our supplies of usable energy. However, many of these advances are currently not economically feasible to implement. It is very important that any energy consumer, whether it be an individual residential consumer or an industrial energy manager, be able to calculate the actual investment requirements and the economic benefits of any proposed scheme, to be able to evaluate more completely its true worth. An economic analysis permits the potential investor to decide whether a particular proposal is economically feasible: for example, whether to trade in an older car that gets only 8 mpg for a new one rated at 35 mpg, or to decide between several options that may achieve the same goal (the comparison of in-situ versus surface retorting of oil shale).

Methods for Profitability Evaluation

Profitability is the general term used to describe the amount of profit that can be obtained from a given situation. The total profit alone cannot be used as the deciding factor in determining profitability. The cost of the project must also be taken into consideration. For example, an industry can make two equally sound investments. The first will cost $25,000 and will save $3000 per year in energy costs. The second will cost $100,000 and will save $5000 per year. Clearly, the savings of $5000 is greater than that of $3000, but the annual rate of return on the second investment is only 5%. [($5000/$100,000) × 100 = 5%], whereas for the first it is 12%. Economically, it would be more advantageous to invest $25,000 in the first option, and invest the difference ($100,000 − $25,000 = $75,000) in, say, bonds or government notes, which might pay around 10% interest.

There are a number of different methods that can be used to determine the profitability of any proposed investment. The following are several of the more widely used techniques.

Return on Investment.   The rate of return on investment (ROI) is calculated, as mentioned above, by dividing the yearly profit by the total initial investment, then multiplying by 100 to convert the fractional return to percent.

$$ROI = \frac{annual\ profit}{initial\ investment} \times 100$$

For an industry, profit is defined as the difference between income (revenue) and expenses. The expenses, or costs, are determined by the capital investment, energy costs, raw material costs, the efficiency of the production process, and similar considerations.

For an individual consumer the profit is the savings attained by making a specific investment. For example, a particular building can be caulked for an investment of $90. The caulking will save $50 per year on heating and cooling costs. The return on the investment is

$$\frac{\$50}{\$90} \times 100 = 56\%$$

There is no single acceptable minimum percentage. Many factors come into play, both for an industry and a consumer. A very low percentage is acceptable if, for example, it is an improvement which, if not made, would result in the industry being closed down. (These issues do not often occur in the energy conservation area). An investment, unless it is required by circumstances such as these, is generally considered advantageous when the return is greater than could be obtained by investing the capital in other ventures: bonds, stocks, even depositing the money in a savings account. Often, however, the ROI analysis is made to compare two competing investments, both of which will accomplish the same overall purpose. In that case the one that generates the greater return on the investment is preferred.

For the individual consumer, the economic feasibility of a particular investment is frequently determined only by comparing the ROI due to energy savings of a particular project with the income obtainable from other investments. However, the consumer may also wish to, for example, replace his or her car, not only for energy conservation reasons, but because the repair bills are too high, the person wishes a more dependable vehicle, or similar. In this case, energy conservation will be of secondary importance. In fact, the overall economic feasibility, the profitability, may be of little importance compared to issues such as style, color, convenience, luxury, dependability, or prestige.

Payback Period.   The payback period, payout period, or payout time is defined as the minimum length of time theoretically necessary to recover the original investment (by savings gained or profit earned):

$$\text{payback period (in years)} = \frac{\text{original investment}}{\text{average annual savings}}$$

The payback period is simply another way of expressing the return on investment. If, for example, the ROI is 25% annually, the payback period is 4 years (1/0.25).

Discounted Cash Flows.   Both of the profitability measures described above are useful, but they do have one major shortcoming: They do not take into account the *time value of money*. If a particular investment were not made, that money would have been available for a second, different investment. This possibility effectively lowers the economic benefits of the first investment, and should be taken into account.

To understand this concept, first consider how interest on a particular investment is calculated. There are a number of different ways that interest can be paid. If money is deposited in a savings account, interest is paid at predetermined intervals. If this interest is added to that already in the account and thereafter earns interest itself, the interest is called *compound interest*. The original deposit is the *principal*, and the interval between interest payments is referred to as the *interest period*. The amount in the account at any given time can be calculated by

$$A = P\left(\frac{1+r}{m}\right)^{mt}$$

where P = principal
   r = interest rate per year (as a fraction)
   m = number of interest periods per year
   t = number of years

Often, interest is paid *continuously*. If, in the formula above, m becomes very large (that is, the interest is calculated and added to the account many times per day), the formula above can be approximated by

$$A = Pe^{rt}$$

This formula is used to calculate the amount in an account if the interest is "compounded continuously."

These formulas for periodic and continuous interest can be inverted and used to calculate the *present value* of money which is forthcoming in the future. Assume, for example, that a person is expecting to be given $5000 two years from now, and that he can earn 10% (continuous compounding) on any money he now deposits in the bank. That $5000 he will get in 2 years' time is equivalent to the $4095 he would receive now.

$$P = Ae^{-rt}$$
$$= 5000e^{-(0.10)(2)} = \$4095$$

That is, if he would have $4095 now, he could invest it at 10%, compounded continuously, and after 2 years the amount in the bank would have accumulated to $5000.

The same principle can be used to analyze the profitability of different projects. Most frequently the time value of money is accounted for by calculating a *discounted cash flow–return on investment* (DCF-ROI). The DCF-ROI can be interpreted to be the maximum interest rate, after taxes, at which money could be borrowed to finance the project where the net savings over the project's life would be sufficient to pay all principal and interest charges accumulated. Or, to put it another way, it is the rate that makes the investment equal to the sum of the present values of all projected future savings.

There are several other concepts that should be considered. Taxes on industrial profits are around 48 to 50%. However, the value of any capital investment may be *depreciated* every year. Depreciation is the decrease in value of physical assets due to physical deterioration, technological advances, or economic changes. The economic function of depreciation can, therefore, be used to distribute (for tax purposes) the original expense over the period that asset is being used. If, for example, a piece of equipment originally cost $11,000, but after 10 years it was worth only $1000 as scrap, the decrease in value would be $10,000. This $10,000 could, for example, be spread evenly over the 10-year life. This additional $1000 expense every year for 10 years could be used for tax purposes.

To see how an economic analysis is made, consider the following. Should a person trade in her 3-year-old car, which gets 8 mpg, for a new $9000 energy-efficient model which averages 35 mpg? The car is driven an average of 12,000 miles per year. Gasoline sells for $1.25 a gallon. She would expect to keep the car 5 years.

The average yearly gasoline savings is

$$\frac{12{,}000 \text{ miles/year}}{(35 - 8) \text{ miles/gallon}} = 444 \text{ gallons/year}$$

$$444 \text{ gallons} \times \$1.25/\text{gallon} = \$556/\text{year}$$

She would pay $9000 in one lump sum at the beginning of the 5-year period to purchase a car. The $556 annual gasoline savings would be spread uniformly over each year of the 5-year period. The present value of the $556 can be calculated by $P = Ae^{-rt}$, where r is the interest rate, t the year, and A the annual savings. Assume that $r = 10\%$; Table 17-1 summarizes the values of e^{-rt}.

Using these values, the present value of the savings can be calculated for each of the 5 years by multiplying the $556 by the appropriate value of $e^{-0.10t}$. These results are shown in Table 17-2.

If the savings on the purchase of gasoline would equal a net 10% of the investment, approximately that which could be earned if the $9000 were invested in other ways, the total or net present value should have equaled the original investment, $9000. Clearly, the $2187.80 is far less than the cost of the car. The purchase would never, in fact, yield a net return on the investment. Even if the purchaser would expect 0% return on her investment, the cost of the car ($9000) would far exceed the total potential savings due to increased gasoline mileage ($2780).

The same type of analysis can be made for an industrial project. In this case, however, the situation is more complex, for operating costs, depreciation, and taxes must also be accounted for.

As an example, consider a pulp and paper company that wishes to begin recycling its process water. There are several advantages to recycling, not the least of which is increased energy savings. Typically in the papermaking process, the water must be heated about 40°F before use. By reusing warm process water, approximately 330 Btu can be saved for every gallon. Depending on the price of fuel, this can result in savings of about $0.66 for every 1000 gallons of fresh water not used.

Assume that the company uses 10,000 gallons of water for every ton of paper produced. The production rate of the plant is 500 tons per day of paper. The capital investment for the recycling is $4,500,000,

TABLE 17-1 Values of $e^{-0.10t}$			TABLE 17-2 Present Value of Gasoline Savings	
Year	$e^{-0.10t}$		Year	Present Value
1	0.9516		1	$ 529.10
2	0.8611		2	478.80
3	0.7791		3	433.20
4	0.7050		4	392.00
5	0.6379		5	354.70
				$2187.80

mainly in plumbing and extra treatment costs. Additional operating costs are $250,000 per year. What is the DCF-ROI for this process change? Assume 330 days of production per year, a project lifetime of 10 years, and a value of zero at the end of the 10 years.

The first step is to calculate the annual energy savings:

10,000 gal/ton of paper $\times$ 500 tons of paper/day $\times$ 330 days/year
$$= 1.65 \times 10^9 \text{ gal/year}$$

$$1.65 \times 10^9 \text{ gal/year} \times \$0.66/1000 \text{ gal} = \$1.09 \times 10^6/\text{year}$$

$$\$1.09 \times 10^6/\text{year} - \$250,000/\text{year} = \$840,000 \text{ savings/year}$$

The depreciation of the $4,500,000 investment can be distributed equally over the 10-year period:

$$\frac{\$4,500,000}{10 \text{ years}} = \$450,000/\text{year}$$

Assume that the construction of the project takes one year. After that initial year, an energy saving of $840,000 per year is experienced. To compute the DCF-ROI, it is necessary to calculate the present value of each of those annual savings. Table 17-3 summarizes values of e^{-rt} which can be used to calculate the present value in accordance with $P = Ae^{-rt}$. The cash flow can be summarized as shown in Table 17-4. A negative amount is enclosed in parentheses.

To calculate the DCF-ROI, it is necessary to estimate what you think might be the annual percentage return, for example, 5% and 10%.

The DCF-ROI corresponding to the project is the percentage that gives a total value of 0 in the last columns. As can be observed from Table 17-5, the return on this project, given these assumptions, is a

TABLE 17-3   Factors with Continuous Interest to Give Present Worths for Cash Flows that Occur Uniformly over One-Year Periods after the Reference Point

Year (t)/r	1%	5%	10%	15%	20%	25%	30%
0-1	0.9950	0.9754	0.9516	0.9286	0.9063	0.8848	0.8640
1-2	0.9851	0.9278	0.8611	0.7993	0.7421	0.6891	0.6400
2-3	0.9753	0.8826	0.7791	0.6879	0.6075	0.5367	0.4741
3-4	0.9656	0.8395	0.7050	0.5921	0.4974	0.4179	0.3513
4-5	0.9560	0.7986	0.6379	0.5096	0.4072	0.3255	0.2602
5-6	0.9465	0.7596	0.5772	0.4386	0.3334	0.2535	0.1928
6-7	0.9371	0.7226	0.5223	0.3775	0.2730	0.1974	0.1428
7-8	0.9278	0.6874	0.4726	0.3250	0.2235	0.1538	0.1058
8-9	0.9185	0.6538	0.4276	0.2797	0.1830	0.1197	0.0784
9-10	0.9094	0.6219	0.3869	0.2407	0.1498	0.0933	0.0581

TABLE 17-4 Cash Flows ($\times 10^4$)

Cash Flows	Years								
	0-1	2-3	3-4	4-5	5-6	6-7	7-8	8-9	9-10
Operating revenues	0	109	109	109	109	109	109	109	109
Operating expenses	0	(25)	(25)	(25)	(25)	(25)	(25)	(25)	(25)
Net from operations	0	84	84	84	84	84	84	84	84
Depreciation	0	(45)	(45)	(45)	(45)	(45)	(45)	(45)	(45)
Pretax profit at 50%	0	39	39	39	39	39	39	39	39
After-tax profit	0	19.5	19.5	19.5	19.5	19.5	19.5	19.5	19.5
Cash flow*	0	64.5	64.5	64.5	64.5	64.5	64.5	64.5	64.5
Net capital	(450)	0	0	0	0	0	0	0	0

*Cash flow = after-tax profit + depreciation.

TABLE 17-5   Calculation of DCF-ROI

Year	Cash Flows (10^4)	Factors (Table 17-3) 5%	Factors (Table 17-3) 10%	Factors × Cash Flow 5%	Factors × Cash Flow 10%
0-1	(450)	0.9754	0.9516	(439)	(428)
1-2	64.5	0.9278	0.8611	59.8	55.5
2-3	64.5	0.8826	0.7791	56.9	50.3
3-4	64.5	0.8395	0.7050	54.1	45.0
4-5	64.5	0.7986	0.6379	51.5	41.1
5-6	64.5	0.7596	0.5772	49.0	37.2
6-7	64.5	0.7226	0.5223	46.6	33.7
7-8	64.5	0.6874	0.4726	44.3	30.5
8-9	64.5	0.6538	0.4276	42.2	27.4
9-10	64.5	0.6219	0.3869	40.1	25.0
				5.5	(82.3)

little greater than 5% (5% gives 5.5; 10% gives −82.3). A more accurate value can be found by interpolating:

$$\frac{5.5 - (-82.3)}{10\% - 5\%} = \frac{87.8}{5\%} = 17.56\%$$

$$(1\%)\left(\frac{5.5}{17.56}\right) = 0.31\%$$

$$5\% + 0.31\% = 5.3\% \text{ DCF-ROI}$$

Under these circumstances, should this industry make this investment? If there are no other circumstances to consider, probably not. Higher interest can be earned elsewhere. But when one considers raw material savings, environmental regulations, and other factors, it may be a very profitable venture.

ENERGY POLICY

There are two major tools of policymaking, particularly regarding the allocation of scarce resources. They can use either market forces or government regulations.

How much do increased prices affect the demand for energy? No one knows for sure, for only minimal research has occurred on such issues. After 1973, however, large computer models were used to make estimates. It was predicted that small price increases of 5% would have no real effect; however, 100% increases would lead to 20 to 30% decreases in use. These predictions have not been borne out in the area of gasoline consumption, where prices have risen over 400%. Although decreases in gasoline consumption have occurred, they have not been as extensive as predicted by the models.

Governmental regulations also pose difficulties. Regardless of the particular policy decisions, one group of people must decide for others regarding their preferences and needs. Prior to 1973, over 70 federal agencies were concerned with various aspects of energy policy. These included the Interior Department, the Oil Import Administration, and the Federal Power Commission, which governed energy development, import restrictions, and price regulation, respectively. Often the regulations clashed. Tax incentives encouraged exploration and development, while price controls on natural gas and oil discouraged investment for these resources. Although much of the government control is now centralized in the Department of Energy (DOE), some of these contradictions may still exist. For example, until early 1981, refiners of domestic oil paid a fee of $5 per barrel, which was used to subsidize the refining of imported oil.

A large number of energy laws were passed in the United States in the years between 1973 and 1980. Table 17-6 is a list of some of these. During this time, the Department of Energy, which began in 1973 as the Federal Energy Office, grew so fast that by 1980 it had more than 20,000 employees, 100,000 full-time consultants and contractors, and an annual budget of over $10 billion.

Many industries now complain about useless and unnecessary regulations. In 1980, for example, the petroleum industry had to file a total of 940,000 DOE forms yearly and spend $400 to $500 million complying with these regulations. A planned coal-fired plant to be built in southern Utah by a consortium of power companies led by Southern California Edison was canceled due to rising cost estimates after a $5 million 4.5-year-long environmental impact statement study, a $8 million investment in planning for pollutant removal, and an increase in projected construction costs from $500 million to over $8.7 billion.

TABLE 17-6   Energy Laws Passed in the United States,
1973-1980

1973	Emergency Petroleum Allocation Act
1974	Energy Supply and Environmental Coordination Act
1975	Energy Policy and Conservation Act
1977	Emergency Natural Gas Act
1978	National Energy Conservation Act
	Powerplant and Industrial Fuel Use Act
	Public Utility Regulatory Policies Act
	Natural Gas Policy Act
	Energy Tax Act
1979	Emergency Energy Conservation Act
1980	Crude Oil Windfall Profits Tax Act

This plant was to be a model of environmental and pollution control. An Exxon Corporation plan for offshore petroleum and natural gas drilling along California's Santa Barbara Channel finally began to produce oil in the early 1980s after over 11 years of work, eight environmental impact studies, 44 consulting studies, 21 public hearings, and expenditures of over $880 million, not including construction or equipment expenses. The natural gas, however, had to be piped back into the field until further governmental studies were made with respect to its extraction.

What should be done? Most experts believe that market forces, guided by open and wise policy decisions and a well-informed public, will be the best path to follow in handling future energy policy questions.

QUESTIONS

1. Summarize how energy prices affect world economies.
2. How do social conditions affect energy prices? Give some specific examples.
3. Research the oil companies' profit in recent years by consulting annual reports and the like. What percentage of their investments are their profits? In what type of operation(s) do the majority of the profits arise?
4. Calculate the payback periods for retrofitting insulation in the ceiling or in the walls of a house. Consult with an insulation supplier. How much would the insulation cost? What type of energy savings would be expected? Which of these is a better investment? List all assumptions.
5. An industry wishes to install a new "hog fuel" boiler, which will burn very cheap fuel. The boiler will cost $800,000, but the savings in fuel costs are expected to be $200,000 per year. For tax and depreciation purposes it is calculated that the boiler will last for 10 years. Assume a 50% tax rate. What is the DCF-ROI? On the basis of energy savings alone, is this a good investment? Why?
6. How has energy policy changed since 1980? What are the major reasons for this change? Consider both economic and political factors.
7. Describe how prices for oil are established in international trade. What is the influence of the spot market?

CHAPTER 18

The Future:
Boon or Bane?

HOW DO WE LOOK INTO THE FUTURE?

The future is an elusive topic for study because it does not now exist. Predicting the future, therefore, does not lend itself to the methods associated with an exact science. This is especially frustrating for scientists engaged in research on energy and the future. For the scientist, there is always the temptation to view the future not from the perspective of "what will it be?" but rather "what do we want it to be?" This was particularly true during the middle of the twentieth century. Much literature that was disseminated to the general public projected a future where all our problems are solved by science and technology. The following examples illustrate this point.

Kirtley F. Mather made the following observations[1] just prior to World War II in an address to the science fraternity of Sigma Xi.

> Enough bituminous and sub-bituminous coal is known to be available within the United States to meet the present annual demand for coal, plus enough to manufacture gasoline and fuel oil in sufficient quantity to meet current demands for at least 2000 years. . . .
>
> Again comes the insistent question from the pessimistic critic. Is there land

[1] Kirtley F. Mather, "The Future of Man as an Inhabitant of the Earth," *Sigma Xi Quarterly*, 1940.

enough? Is there sufficient fertile soil to provide adequate food and, in addition, the plant material for the ever expanding chemical industries? And again we hear the same reply. Yes, there is enough and to spare. . . .

There is, therefore, no prospect of the imminent exhaustion of any of the essential raw materials, so far as the world as a whole is concerned, provided our demands for them are not multiplied rapidly in the future. . . .

Taking all these things into consideration, it would appear that the world stores of needed natural resources are adequate to supply a basis for the comfortable existence of every human being who is likely to dwell anywhere on the face of the earth for something like a thousand years to come. . . .

The unbridled optimism of forecasters had not diminished about 25 years later, in 1957, when H. Russell Austin, a writer for the *Milwaukee Journal*, predicted what life would be like in the year 1982. Some of Austin's prophecies[2] include the following:

Vertically rising and landing planes requiring less airport room. . . .

A rocket mail letter from Milwaukee to New York takes only 20 minutes, with hourly launchings. . . .

Travel overseas by huge, 2000 mile an hour jet airliners, however, has become relatively inexpensive, and families of average income often make such trips during their annual vacations. . . .

Some neighborhoods . . . are supplied with current from a small underground atomic reactor. . . .

They . . . enjoy a living standard and annual purchasing power about 65% higher than in the 1960's. . . .

Mechanization of the farms has gone further than the boldest dream of 35 years ago: The electronically guided unmanned tractor which plows, fertilizes, cultivates, or harvests by drawing various attachments in a predetermined but variable pattern; the automatic, air conditioned dairy barn, which cares for, feeds, and milks the cows at preset hours, then turns them out to pasture, cleans itself, pasteurizes and cans the milk for shipment; and the automatic chemically fed vegetable garden, with temperature and moisture regulated year round under transparent plastic roof, and with automatic machine harvest are among the everyday wonders of agriculture. . . .

The forecasts we have looked at are extensions or extrapolations of what currently exists, assuming that the best of what we have will be maintained and expanded and that the problems prevalent at that time really stay the same or diminish, such as the economic gulf that exists between developed and nondeveloped countries of the world.

Time also allows changes to take place. Sometimes these changes

[2] H. Russell Austin, "The Wisconsin Story: Building of a Vanguard State," *Milwaukee Journal*, 1957, pp. 457–464.

occur through an orderly process of deliberate steps. However, it is often difficult to predict the forces that motivate the change. Nuclear technology developed rapidly on a crash program during World War II. Rapid implementation of existing technology occurred in energy conservation efforts when gasoline and fuel oil supplies were no longer inexpensive and dependable. Not all world problems, however, generate a "crisis" state of mind. There are more than 12,000,000 people who die each year from starvation, yet the worldwide efforts to deal with this problem are not improving the situation.

The Council of Environmental Quality prepared an estimate in 1980 of world conditions in the year 2000.[3] A summary of that report includes the following points:

1. On a worldwide scale, the people of the world will probably be poorer than they are today.
2. The world will be more crowded, with an increase from 4.25 billion to 6.3 billion people. About 90% of the increase will come from the poorest countries.
3. Food production will increase about 90%, which, with the increase in population, amounts to about 15% more food per person, but with a disproportionate amount of the increase going to the people who already have enough food. Some populations may actually have less to eat at the end of the century than they have now.
4. The increase in food production will result from more intensive agriculture, which will require much larger energy inputs.
5. One-fourth of the world will continue to use three-fourths of the world's mineral resources.

Because energy resources will continue to be expensive, it is becoming increasingly important that we learn how to project world energy demands much more accurately than we have in the past. Human needs determine the demand for energy. But "needs" must now be defined in terms of income and the cost of energy. Simply knowing how many clothes dryers are being sold no longer tells us how much energy they will need, because some people will use them more frequently than others. We must also factor in different climates and local weather patterns. An increasingly important question is not only will the dryer be used, and how much will the dryer be used, but *when* will it be used? Time-of-day pricing for electricity can affect the "when," resulting in a leveling of the daily peaks and valleys in energy use. This, in turn, increases the efficiency of equipment use.

[3] *Entering the Twenty-First Century: The Global 2000 Report to the President*, The Council on Environmental Quality and the Department of State, 1980.

Figures 18-1 and 18-2 illustrate some of the problems associated with projecting energy demand by simple extrapolation or just "extending the line." How much money we will have to spend and how we will choose to spend it keeps the "experts" guessing, whether they are Madison Avenue advertisers or theoretical economists who live in ivory

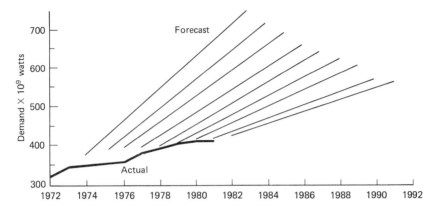

Figure 18-1 Projecting peak demand for electric energy. (Courtesy of the Electric Power Research Institute.)

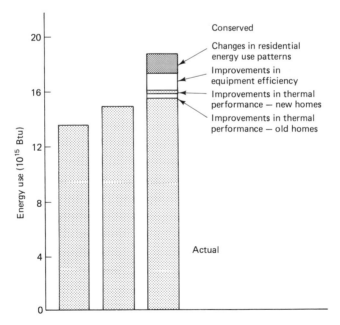

Figure 18-2 Residential electrical energy use. (Courtesy of the Electric Power Research Institute *EPRI Journal*, vol. 7, no. 10, December 1982, p. 11.)

towers. New modeling techniques must be more comprehensive than those in the past. If the projection is too high, valuable money resources are spent on equipment that is not fully utilized and therefore may not pay for itself, let alone provide a profit. If the projection is too low, there is no inexpensive way to catch up.

Scenarios represent an alternative method for estimating future needs. A scenario is a series of events that we imagine might happen in the future. We begin by posing a series of "what if" questions. If a scenario is to be effective, it must be reasonable accurate, but it must also project an appropriate attitude or philosophy regarding the future. All the scenarios we have looked at so far have assumed that the quality of life will be significantly better for everyone. This perspective reduces the effectiveness of a scenario because we are being lulled into a false sense of security that tells us that no action on our part at this time is really necessary. Scenarios with a very negative perspective can be just as useless, because a sense of doom may lead to paralysis or inaction.

Hawkins, Oglivy, and Schwartz[4] have developed seven scenarios that project a wide variety of possibilities. One, the "Official Future," which takes on several different forms in an election year, assumes that the world is basically in good shape and that the pursuit of material abundance is still the main driving force for change. Little emphasis is put on the interrelatedness of activities and their consequences and individuals assume that their specific activities are really rather meaningless in the larger scheme of things. Tomorrow will be much like today, only better.

Other models of the future consider the possibility that things could begin or, as some perceive it, continue to go wrong. Several answers come to mind. We could develop a more authoritarian form of government where dissent is not tolerated. We could move toward becoming "leaner and wiser." We could have a near miss with a calamity that is recognized by everyone, such as a nuclear holocaust, and quickly "see the light and error of our ways."

But what if things simply go from bad to worse? We could suffer from a decay of imagination and be forced to turn to less secular directions for answers. The ultimate doom is when the stubborn resistance to events gives way to panic: We face our severest test and fail. Under these conditions, the world degenerates into thousands of local economies. We are back at the beginning, but there is nothing left with which to start over. There is yet another possibility. If the system does break down, people may recognize that they are the really important resources and begin to help each other.

[4] Paul Hawkins, James Oglivy, and Peter Schwartz, *Seven Tomorrows* (New York: Bantam Books, Inc., 1982).

The real answer lies in our perception of ourselves. The day after the stock market crash in 1929 we still had the same quantity of physical resources that existed the day before. We had not experienced an invasion from the outside; rather, we experienced a change in the perception we had of ourselves and what we could accomplish. We react to a world that we believe is there. This may be our greatest asset or our greatest liability.

THE FUTURE

What is a reasonable estimate of the future regarding energy sources and consumption? A number of different study groups have dealt with this and predict, quite often, diversity in both overall thrust as well as "details." However, there is some agreement about a few combinations of technological innovations and life-style changes that will probably occur. The Electric Power Research Institute stated in January 1981: "The issue is not a choice of conservation *or* coal *or* nuclear *or* renewable resources, rather it is our willingness to use conservation *and* coal *and* nuclear *and* renewable resources. We do not enjoy the luxury of selecting from these alternatives; we must utilize all of them."[5]

The large-scale development of new sources of energy could reduce our need to conserve. Coal and nuclear (fission) power are, however, generally considered to be the only feasible short-term supply alternatives that might have a significant impact on our energy demand, although some experts now predict that natural gas may be in adequate supply to have a significant impact in the immediate future. Solar, biomass, and fusion energy, if they can be economically and technologically perfected, show promise for the more distant future. Regardless of their feasibility and application, however, it is clear that we cannot continue increasing our rate of energy consumption as we had been for the past several decades. Our energy consumption curve must continue to level off, even decrease, in the near future if we wish to avoid serious energy supply problems.

Reduction in Energy Use

Projections of our future energy consumption continue to be modified from those originally made in 1972. For example, in 1979 the U.S. Department of Energy predicted that the U.S. energy demand would reach 122 quads by 2000. By 1981, that projection had been

[5] "Electricity Shortage Possible before 2000," *Chemical and Engineering News*, January 5, 1981, p. 6.

TABLE 18-1   1981 Estimates of U.S. Energy
Consumption for the Year 2000

	Quads
U.S. Solar Energy Research Institute	62–66
University of Michigan-Princeton University	64
National Audubon Society	80
Mellon Institute	88
U.S. Department of Energy	102
Edison Electric Institute	117

decreased to 102 quads. Other groups and agencies varied widely in their projections, as shown in Table 18-1.

How are these reductions in predicted energy consumption likely to come about? Let us consider several possible impacts on future energy use.

Housing Patterns.   It is predicted by some that more and more people will be living in multifamily dwellings such as condominiums, apartment houses, and duplexes. This will appreciably reduce heating costs as well as, in many cases, make the homes affordable given today's rapidly rising construction costs.

More people will also be living in intermediate-size cities: A city of 100,000 is much easier and less expensive to get into and out of, particularly by auto, than is a city of several million. The shift of population to the South and West will probably continue, necessitated in part by high energy costs.

Shopping Patterns.   It is like that the growth of shopping malls, located on the outskirts of larger cities, will continue, because for most residents they are much more accessible than the traditional downtown stores. These malls will include grocers, clothiers, restaurants, druggists—almost everything that is required to satisfy the local residents' needs.

Transportation.   It is predicted that people will be driving shorter distances due to a change in settlement patterns. In the United States we drive 8000 miles per year per capita. People will probably be living closer to their work and to stores. Safe walking and biking will further reduce the necessity for driving. Currently, 25 to 50% of many urban areas are used to store autos. This could be reduced if fewer people drove private autos to do their shopping. Simultaneously, there would be less noise and fewer exhaust emissions. Another way that travel will be reduced is by the substitution of improved telecommunications for travel. This is particularly true for business-related travel.

There will be a heavy emphasis on rapid rail systems for intercity travel, especially for distances greater than 500 miles. The United States currently lags behind many European countries in the development of fast rail service, but improvements in this area will occur. Cars and buses will still be important, however.

The cars of the future will be smaller, lighter, and (since the energy performance of an auto is approximately proportional to its weight) more fuel efficient. The government mandated that average gasoline consumption by cars be 27.5 mpg by 1985. It is likely the cars will also be used longer (less planned obsolescence), perhaps up to 20 years, or 200,000 miles.

Fuels will also change. Currently, 54% of our petroleum is used for transportation. It would be preferable if we could supply at least part of our demand with another, renewable fuel.

Industrial Production.   There will probably be a twofold approach to industrial energy consumption: (1) Industries will switch to the most energy-efficient process technologies and will continue to implement cogeneration and other energy-savings methods, and (2) simultaneously there will be a change in the mix of goods and services. Manufactured goods will last longer due to an increase in durability. Automobile production will probably decrease due to both decreased ownership and less planned obsolescence. Recycling will help extend our energy and other resources.

Food Production.   Although food production will not decrease, there will be a shift in the types of foods that are eaten. Energy-intensive foods such as feedlot beef will decline in popularity. Meanwhile, farms will become less specialized; more local crops will be consumed, reducing transportation costs, and there will be growth in home (vegetable) gardening and home orchards.

New Energy Technologies

To predict better the changes in energy sources that are likely in the future, it is useful to consider the modifications that have already occurred in the period immediately following the 1973 oil embargo. Figure 18-3 depicts the changes in domestic energy production both in the pre- and post-embargo periods. Increased coal and nuclear power consumption had more than offset the decline in production of natural gas and petroleum during the period 1973–1980; furthermore, these two sources supplied 98% of the growth in domestically produced

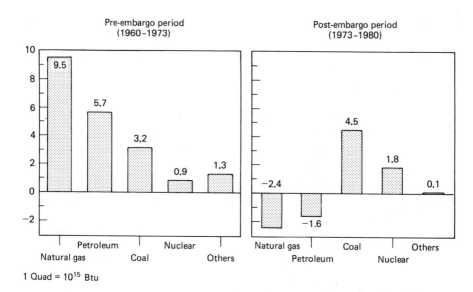

1 Quad = 10^{15} Btu

Figure 18-3 Changes in domestic energy production, pre- and post-embargo periods. (Courtesy of the Westinghouse Electric Corporation.)

energy in this 7-year period. As discussed in Chapter 8, however, the discovery of new, large deposits of natural gas have led to a resurgence of its use since 1980, and show promise that it may remain an important fuel into the next century.

What about the new technologies? Why haven't they had more of an impact? Will they have a major influence in the near future? To answer this question, it is important to understand the stages involved in developing an idea from the laboratory stage to commercialization.

Technological Development.   Whenever a new technology appears on the market, it typically passes through four stages of development. The basic idea generally arises out of knowledge gained by basic and/or applied research conducted in a scientific laboratory. If the technology is not scientifically feasible, it will never be feasible. Some research can occur rapidly, and sudden breakthroughs are always a possibility, but they are rare. Generally, this stage is very time consuming, even tedious.

The next stage is to demonstrate the engineering feasibility. Is the process actually going to work on a large scale? Usually, a pilot plant is constructed. The pilot plant incorporates all important subsystems, whose performance can then be checked. If that is successful, a commercial-scale plant is constructed. Issues such as cost, reliability, and product quality can then be evaluated.

If the foregoing two stages are successful, commercialization begins. Private investors begin to place orders for the new equipment or the process itself. This may occur quite slowly. New plants require time to build and it takes time before old plants are truly obsolete. If the new technology requires significantly different methods of operation and maintenance, these support services also need time to become established.

The last stage, if the technology is to be successful, is to prove its economic attractiveness. If a technology is brought to the market before all of its potential problems are worked out, it may be rejected; on the other hand, if development is delayed too long, newer and better technologies will make it obsolete before it becomes established.

The Time Lag of Energy Innovation. What time period are we discussing when we consider bringing a technology from the laboratory to successful commercialization? This usually depends on the nature of the technology. Table 18-2 summarizes the time lag for a number of technologies. Note that this time lag has been continually decreasing.

Why, then, do scientists and industrialists predict that most of the alternative energy technologies won't have a significant impact for several more decades? The answer is inherent in the nature of electric power generation, since most of these new technologies are designed to produce electricity.

At the present time, 35 to 45 years are needed to bring a new electric power technology from the laboratory to large-scale commercial use, as witnessed by the introduction of fission power. The large scale

TABLE 18-2   Time Lag for Technological Innovations

Scientific Discovery	Practical Application
Steam	160 years from the work of Huygens to the first steam railroad
Electricity	80 years from the discoveries of Ampère and Volta to the first power plant
Wireless transmission	60 years from the studies of Maxwell to the first television set
Atomic power	50 years from theories of Planck and Einstein to the first nuclear-powered ship
Lasers	7 years from the theory to medical, scientific, and industrial applications
Transistors	6 years from development in Bell Laboratories to use in radios

of the electric power industry certainly precludes rapid change. But other factors are just as important.

1. The side effects must be carefully monitored to prevent environmental damage. Even for coal-fired plants, 4 to 5 years is required to obtain the necessary governmental approvals before a new facility can be built.
2. Public commitment to any single technological alternative is unlikely. A small supply of money will necessarily be split into many fractions.
3. New technologies such as fusion are more complex, requiring more sophisticated equipment, than the current technologies require. This, in turn, creates a need for highly trained personnel, hence additional time.

What would happen if an unlimited supply of money were available? The time frame would probably not be altered very much. In the case of fusion, large budgets have ensured that several promising alternatives to fusion containment could be investigated. But for other technologies, the only shortcuts that would be possible, such as going directly from the laboratory stage to commercialization, might be very detrimental in the long run. A prime example is the heat pump. Heat pumps were first introduced in the 1950s. Their supporters claimed that they were more economical, in spite of higher capital cost, because of their high efficiency. Unfortunately, many of the early heat pumps were plagued by mechanical problems. Sales declined rapidly, and only after another two decades did they again receive serious attention as a feasible heating method.

Solar energy technology can be used to illustrate these points. Solar energy consists of two separate types of technologies. Each involves a different set of delays. Although solar units are now available for domestic hot-water and space heating, the high capital investment for the consumer has slowed their adoption. Consumer financing is only now being developed, and the long-term reliability of these units is still a question. In addition, solar access rights have been legalized by relatively few municipalities. But even if solar installations were deemed feasible by a large share of our population, the slow turnover in housing would delay their impact. Even if 25% of the new homes built in the remainder of this century included solar space heating and 3.5 million homes were retrofitted, the savings would be only about 1% of total U.S. energy requirements.

The technology for utility-sized solar production of electricity is not yet thoroughly developed. It is still at the pilot-plant stage. The eventual success of solar-generated electrical energy depends on tests now under way. It is unlikely that commercialization could begin before the 1990s. Thus it is unlikely that solar power could make a major contribution to our electric generating capacity before the next century.

A BACKWARD GLANCE

Our generation often feels that it is the only one wise enough to identify the problems that face us, and ingenious enough to suggest possible solutions. Lest we become overconfident, let us take a glance at three paragraphs printed in *Scientific American* in the mid-1920s.

December 1924:

There is suspicion that our resources of coal and oil are being wasted, that their exhaustion is no more than around the corner of the next century, and that our civilization is threatened, in consequence, with an early and disastrous end. There is talk of possible power from the waves and the winds, and of the stores of power that some scientists believe may be obtainable from the atoms of matter. Before considering these alternatives, however, we must first take stock of what resources we have. That was done during the past summer at World Power Conference in London. Hundreds of engineers and experts met to report on the power resources of their respective countries. The conference resulted, among other things, in the most complete and accurate survey of world power resources ever made. The total coal of the world, for example, was found to aggregate 7,397,553,000,000 metric tons. At the present rate of use this is enough for some 4,000 years. Of course, the rate of use is increasing.[6]

August 1926:

When President Coolidge appointed the Federal Oil Conservation Board, he wrote a letter that showed what a thorough grasp he had obtained of the fundamental facts governing the demand and supply of oil. The letter opens by stating, "It is evident that the present methods of capturing oil deposits is wasteful to an alarming degree." In another paragraph he says that he is advised that our current oil supply is kept up only by drilling many thousands of new wells each year, and that "the failure to bring in producing wells for a period of two years would slow down the wheels of industry and bring about a serious industrial depression." Julian D. Sears of the United States Geological Survey reminds us that it required 41 years and four months to produce the first billion barrels of oil, that it took only eight years and one month to bring in another billion barrels and that only one year and seven months was

---

[6] *Scientific American*, Vol. 131, No. 6, December 1924.

required for the seventh billion to be brought to the surface. In the presence of this rate of increase anyone who tells the American people that the cistern has no bottom is doing the country a great disservice.[7]

January 1928:

In spite of the many warnings by oil men who speak with the voice of knowledge and authority, the wild and greedy scramble to suck our rich oil deposits up to the surface and turn them into cash goes on apace. Our surpassing wealth today is to be credited, more than anything else, to the lavish way in which nature has endowed the United States with valuable resources. Both the Federal Government and the wisest leaders in the oil industry are striving to find some way to check the prevailing extravagance. Secretary Work of the Department of the Interior has suggested the creation of a committee of nine to study legislative action, and both the American Bar Association and the oil industry have appointed their respective members. The Oil Conservation Board of the Federal Government now will name three representatives. Note, if you please, that this call by the Government is for the purpose of studying "legislative action." We have long been of the opinion that only by legislative action can the present orgy of greed and wastefulness be stopped. Appeals to the good business sense and to the patriotic feelings of the oil getters alike seem to have fallen on deaf ears. There is no indication of a willingness among the personnel of the oil industry at large to get together for the purpose of controlled drilling. It is high time for the law to step in.[8]

Although both scientists and government officials recognized many years ago some of the upcoming problems, little was done to prevent the "crisis" we faced in 1973. Will we do any better for the future? Only time will tell.

QUESTIONS

1. Kirtley Mather stated in 1940 that there is enough coal available for 2000 years. How did he qualify this statement? If the statement was true, would a similar statement be true today? Explain.

2. Mather states that there is sufficient land available to meet the world food demand. If this statement is true, why are 12,000,000 people starving each year?

3. What general assumptions are being made by many people when they predict that life is going to be better for everyone in the future?

4. How does the observation "The more things change, the more they stay the same" apply to the world economy?

[7] *Scientific American*, Vol. 135, No. 2, August 1926.
[8] *Scientific American*, Vol. 138, No. 1, January 1928.

5. The Council of Environmental Quality predicts that by the year 2000 food production will increase by 90%. What other predictions do they make concerning food that suggest that the world food problem will not be solved?

6. The clothes dryer is cited as an example of a household appliance that can have an energy demand that varies a lot. What other household appliances can be placed in this same category of variable energy demand? What factors might influence the energy demand of each?

7. Why are very favorable or very unfavorable predictions not particularly useful?

8. The stock market crash of 1929 was cited as an example of a sudden change in the attitude in the United States that had little to do with catastrophic physical circumstances. Can you think of other times when the attitude in this country changed dramatically in a short period of time and the triggering event was not a physical calamity affecting large numbers of people?

Energy Relationships

Basic Unit Conversions

English	Metric
1.00 pound of force*	453.6 grams of mass
2.2 pounds of force	1.00 kilograms of mass
1.00 inch	= 2.54 centimeters (cm)
39.37 inches	= 1.00 meter (m)
1.00 quart	= 946 milliliters (ml)
1.0 second	= 1.0 second

*When the acceleration due to gravity remains constant at 9.8 m/sec^2, 453.6 grams of mass experiences a force of 1.00 pound.

Mechanical Energy Relationships

$$\text{Velocity} = \frac{\text{distance}}{\text{time}}$$

$$1.0 \text{ ft/sec} = \frac{1.0 \text{ ft}}{1.0 \text{ sec}} \qquad 1.0 \text{ m/sec} = \frac{1.0 \text{ m}}{1.0 \text{ sec}}$$

$$\text{Acceleration} = \frac{\text{change in velocity}}{\text{time}}$$

$$1 \frac{\text{ft}}{\text{sec}^2} = \frac{1.0 \text{ ft/sec}}{1.0 \text{ sec}}$$

Force = mass × acceleration

$$1.0 \text{ pound} = (1.0 \text{ slug}) \times \left(1.0 \frac{\text{ft}}{\text{sec}^2}\right)$$

$$1.0 \text{ Newton (N)} = (1.0 \text{ kilogram}) \times \left(1.0 \frac{\text{m}}{\text{sec}^2}\right)$$

$$1.0 \text{ dyne} = (1.0 \text{ gram}) \times \left(1.0 \frac{\text{cm}}{\text{sec}^2}\right)$$

Work = force × distance

$$1.0 \text{ ft-lb (foot-pound)} = (1.0 \text{ lb}) \times (1.0 \text{ ft})$$

$$1.0 \text{ joule} = (1.0 \text{ N}) \times (1.0 \text{ m})$$

$$1.0 \text{ erg} = (1.0 \text{ dyne}) \times (1.0 \text{ cm})$$

Heat Energy Relationships

English	Metric
British thermal units	Calorie
1.0 Btu = amount of heat required to change the temperature of 1.0 lb of water $1.0°F$	1.0 cal = amount of heat required to change the temperature of 1.0 g of water $1.0°C$

$$1.0°F = (5/9) \times (1.0°\text{Celsius})$$
$$1.0 \text{ Btu} = 252 \text{ cal}$$
$$1.0 \text{ Btu} = 778 \text{ ft-lb}$$

Energy and Power*

English	Metric
$1.0 \text{ hp (horsepower)} = \dfrac{550 \text{ ft-lb}}{1.0 \text{ sec}}$	$1.0 \text{ watt} = \dfrac{1.0 \text{ joule}}{1.0 \text{ sec}}$

$$\text{Energy} = (\text{power}) \times (\text{time})$$
$$1.0 \text{ watt-hour} = (1.0 \text{ watt}) \times (1.0 \text{ hour})$$
$$1.0 \text{ kWh (kilowatt-hour)} = (1.0 \text{ kilowatt}) \times (1.0 \text{ hour})$$
$$1.0 \text{ kWh} = 3,600,000 \text{ joules}$$
$$1.0 \text{ kWh} = 3140 \text{ Btu}$$

*Power $= \dfrac{\text{energy}}{\text{time}} = $ rate of use of energy.

World Energy Use in Servant Equivalents

United States	80	Argentina	11
Canada	67	Spain	10.0
Czechoslovakia	45	Yugoslavia	9.5
East Germany	42	Chile	9.0
Sweden	42	Mexico	8.3
Australia	40	Cuba	8.0
United Kingdom	39	Greece	7.7
Denmark	36	Taiwan	6.2
West Germany	35	Uruguay	6.2
Norway	33	Lebanon	5.4
Netherlands	31	Iraq	4.9
USSR	31	Colombia	4.4
Bulgaria	26	South Korea	4.4
Finland	26	Brazil	3.4
France	23	Syria	3.4
Switzerland	23	Turkey	3.4
Austria	22	South Vietnam	2.6
Hungary	22	United Arab Republic	2.3
Ireland	22	India	1.6
New Zealand	21	Bolivia	1.5
South Africa	21	Thailand	1.5
Japan	19	Ghana	1.0
Romania	19	Pakistan	0.8
Italy	17	Ethiopia	0.2
Israel	16		
		World average	12.9

Source: U.S. Department of Energy.

APPENDIX **C**

Annual Energy Requirements of Electric Household Appliances under Normal Use

When using these figures for projections, such factors as the size of the specific appliance, the geographic area of use, and individual usage should be taken into consideration. Note that wattages are not additive since all units are normally not in operation at the same time.

Appliance	Average Wattage	Estimated kWh Consumed Annually
Food preparation		
Blender	386	15
Broiler	1,436	100
Carving knife	92	8
Coffee maker	894	106
Deep fryer	1,448	83
Dishwasher	1,201	363
Egg cooker	516	14
Frying pan	1,196	186
Hot plate	1,257	90
Mixer	127	13
Oven, microwave (only)	1,450	190
Range		
With oven	12,200	1,175
With self-cleaning oven	12,200	1,205
Roaster	1,333	205
Sandwich grill	1,161	33
Toaster	1,146	39
Trash Compactor	400	50
Waffle iron	1,116	22
Waste disposer	445	30
Food preservation		
Freezer (15 ft^3)	341	1,195
Freezer (frostless 15 ft^3)	440	1,761
Refrigerator (12 ft^3)	241	728
Refrigerator (frostless 12 ft^3)	321	1,217

Appliance	Average Wattage	Estimated kWh Consumed Annually
Refrigerator/freezer		
14 ft^3	326	1,136
Frostless 14 ft^3	615	1,829
Laundry		
Clothes dryer	4,856	993
Iron (hand)	1,008	144
Washing machine (automatic)	512	103
Washing machine (nonautomatic)	286	76
Water heater	2,475	4,210
Water heater (quick recovery)	4,474	4,811
Comfort conditioning		
Air cleaner	50	216
Air conditioner (room)	860	860*
Bed covering	177	147
Dehumidifier	257	377
Fan (attic)	370	291
Fan (circulating)	88	43
Fan (rollaway)	171	138
Fan (window)	200	170
Heater (portable)	1,322	176
Heating pad	65	10
Humidifier	177	163
Health and beauty		
Germicidal lamp	20	141
Hair dryer	381	14
Heat lamp (infrared)	250	13
Shaver	14	1.8
Sunlamp	279	16
Toothbrush	7	0.5
Vibrator	40	2
Home entertainment		
Radio	71	86
Radio/record player	109	109
Television		
Black and white, tube type	160	350
Black and white, solid state	55	120
Color, tube type	300	660
Color, solid state	200	440
Housewares		
Clock	2	17
Floor polisher	305	15
Sewing machine	75	11
Vacuum cleaner	605	46

*Based on 1000 hours of operation per year. This figure will vary widely depending on area and size of unit.

Source: Electric Energy Association.

Transportation
Energy Conservation
Techniques

Walk instead of drive. Walking 1 mile every day instead of driving saves 80 gallons of gasoline per year and $80 for large cars and 15 to 20 gallons of gasoline for small cars.

Take the bus. Buses are about four times more efficient than cars in terms of the amount of fuel needed to transport each passenger 1 mile.

Join a car pool. With three people per car, the average savings amounts to 150 gallons of gasoline per year. Having six people per car amounts to an average savings of 240 gallons of gasoline per year.

Cut shopping trips to once a week. Average savings equals 20 gallons of gasoline per year.

Walk to work or ride a bicycle. Average savings equals 350 gallons of gasoline per year.

Take a train or bus instead of an airplane. On a long trip this can amount to be a savings of 80 gallons of gasoline per person.

Put a timer on your engine block heater. Running it for 8 hours only can save you 200 kilowatt-hours of electricity and $8.

Plan your shopping trips so that you drive the shortest possible distance.

Household Energy Conservation Techniques

Put up storm windows. Save 150 to 480 gallons of fuel oil each year. This represents saving $135 to $480 each year in heating oil bills.

Use weather stripping. Save 120 to 450 gallons of fuel oil each year. This represents saving $105 to $400 each year in heating oil bills.

Add ceiling and wall insulation. Save 30 to 600 gallons of heating oil each year. This represents saving $27 to $450 each year in heating oil bills.

Close window drapes at night. This can amount to a savings of as much as 80 gallons of fuel oil.

When you replace your furnace, buy an energy-efficient one. An energy-efficient furnace can save as much as 350 gallons of fuel oil each year.

Turn your thermostat down at night. It takes less energy to bring a house back to the desired daytime temperature than to maintain that temperature all night long.

Insulate your ceiling first. From no insulation to 3.5 inches of insulation in a ceiling reduces heat loss by 80%, whereas this same change in the walls reduces heat loss by only 25%.

If you have an air conditioner, plant trees and install awnings on the east and west sides of your house. This can reduce electric energy use by 500 kilowatt-hours. This is a savings of $20 per year.

If you have an air conditioner, raise the inside temperature of your house by 5° F. This will reduce your electric energy usage by as much as 850 kilowatts. Raising the temperature 10° F, will reduce the amount of savings by 700 kilowatts. This is a savings of $15 to $30.

If you have an air conditioner, leave your storm windows on in the summer. This can reduce your electric energy use by 200 kilowatt-hours. This is a savings of $8 to $10.

If you have a room air conditioner, use it in a room that faces north, keep the temperature at 80° F, and close the door. This combination will result in an electrical energy savings of about 700 kilowatts of energy. This is a savings of $28 to $80.

Replace your hot water heater with an energy-efficient one. This can result in an energy savings of 700 kilowatt-hours or 7500 cubic feet of natural gas. This is a savings of $80 or $18, respectively, each year.

Save energy on hot water by installing flow restrictors on shower heads and faucets, turning the water off when you are on vacation, lowering your water temperature by 20° F, and insulating your water pipes. This can reduce your energy use by 1400 kilowatt-hours or 15,000 cubic feet of natural gas. This is a savings of $55 and $38, respectively.

Replace your refrigerator with an energy-efficient model. This will reduce electricity demand by 700 kilowatt-hours and save you $30.

Replace your refrigerator with a smaller model. This will reduce electricity demand by 700 kilowatt-hours and save you $30 each year if you go from a 17-cubic-foot to a 12-cubic-foot model.

Replace your freezer with an energy-efficient model. This will reduce your energy demand by 350 kilowatt-hours and save you $15.

Decide what you want before you open the refrigerator or freezer so that the door is open only a short time. This can reduce your electrical demand by 50 to 200 kilowatt-hours and save you $6 to $8.

Keep your freezer more than half full. This can reduce your electrical demand by 200 kilowatt-hours and save you $8.

Hang your clothes outside instead of using the dryer. This will save 1000 kilowatts of electricity and save you $40. Reducing dryer use to one load per day will save you $20 per year.

Turn off the television set when no one is watching it. For a color set this can save 850 kilowatt-hours each year and you will save $15.

Do not use the dry cycle on the dishwasher. This will reduce the energy used by the dishwasher by 50%. That will be a 200-kilowatt-hour reduction and an $8 savings for you.

Replace your stove with an energy-efficient model. This will reduce your energy use in this area by 25%. That is a savings of 350 kilowatt-hours of electricity and 4000 cubic feet of natural gas. That is a savings of $15 and $10, respectively.

Change one-half of your light bulbs to fluorescent bulbs. This will reduce your electric demand by 350 kilowatt-hours and save $15.

Do not leave lights on all night. A 50-watt bulb burning all night requires 150 kilowatt-hours of electricity each year. This costs you $8.

Clean the coils on your refrigerator and freezers.

Fix leaking hot-water faucets.

Check the seal on your refrigerator door.

Use the right-size pan for cooking.

Use as little electricity as possible at noon and from 4 P.M. to 6 P.M.

Remember that everything you buy requires energy to make and to deliver to you even if it does not require energy to use it.

World Distribution of Nuclear Power Plants

Country	Number of Nuclear Plants	Country	Number of Nuclear Plants
United States	189	Finland	4
France	53	Hungary	4
USSR	41	Argentina	3
United Kingdom	39	Austria	3
Japan	30	Brazil	3
West Germany	28	Mexico	2
Canada	24	Netherlands	2
Spain	16	Philippines	2
Sweden	12	Poland	2
Czechoslovakia	9	Romania	2
Italy	9	South Africa	2
India	8	Egypt	1
Belgium	7	Iraq	1
East Germany	7	Libya	1
Korea	7	Luxembourg	1
Switzerland	7	Pakistan	1
Taiwan	6	Turkey	1
Bulgaria	4	Yugoslavia	1

Source: Alvin Weinberg, "The Future of Nuclear Energy," *Physics Today*, Vol. 3, No. 48, March 1981.

World Coal Reserves, 1977 [Billion (10⁹) Metric Tons]

	Geological Resources				Technically and Economically Recoverable Reserves			
	Hard Coal	Brown Coal	Total	Percent	Hard Coal	Brown Coal	Total	Percent
Africa	173	—	173	(1.7)	34	—	34	(5.3)
North America	1,286	1,400	2,686	(26.6)	122	65	187	(29.4)
Latin America	22	9	31	(0.3)	5	6	11	(1.7)
Asia*	5,494	887	6,381	(63.0)	219	30	249	(39.1)
Australia	214	49	263	(2.6)	18	9	27	(4.2)
Europe	536	54	590	(5.8)	95	34	129	(20.3)
Total	7,725	2,399	10,124	(100)	493	144	637	(100)

*Asia includes European USSR.

Projected Coal Use, 1990 and 2000 (Million Metric Tons)

	1978 Actual	1990		2000	
		CIAB Questionnaire	WOCOL Report	CIAB Questionnaire	WOCOL Report
Electric and heat	597	1194–1231	860–1131	1557–1625	1243–1748
Industry steam and hot water	82	193–198	145–201	308–335	219–378
Iron and steel	205	261	272–279	304–311	297–313
Coal conversion	1	40	10–26	166–187	156–353
Total	884	1688–1730	1287–1637	2335–2459	1915–2792

Index

A

Acceleration, 251
Acid rain, 74-75, 93
Agriculture:
  domestic animals, 59
  energy inputs, 61
  efficiency, 60
  fertilizer, 60
  human labor, 59
Agriculture energy conservation,
    213-15
  biomass, 215
  efficiency, 213
  single-pass agriculture, 214
  solar buildings, 215
  windmills, 215
Agriculture energy use, 59-62
Air pollution:
  combustion, 74
Airplanes, 206-8
Alaskan oil production, 93
Algae, 184
Alternating current, 118-19
Alternative energy sources, 64
Aluminum, 212
Automobile, 50, 205-9
  efficiency, 52
  pollution, 50

B

Baghouses, 78
Base-load plants, 131, 222
Bicycle, 205
Biomass, 182-90
  algae, 184
  buffalo gourd, 185
  Chinese tallow tree, 185
  gopher plant, 185
  kelp, 184
  sunflower oil, 184
  water hyacinths, 184
Black lung disease, 73
Blending agents, 187
Breeder reactor, 54, 141, 154
  locations, 154
British thermal unit (Btu), 252
Buffalo gourd, 185

C

Cable-tool drilling, 94
Calorie, 252
Carbon dioxide:
  greenhouse effect, 26-27
Carrying capacities, 12
Catalylic cracking, 97-98